A Systemic History of the Middle Way

Middle Way Philosophy

Series Editor: **Robert M. Ellis**, Middle Way Society

Middle Way Philosophy is a cross-disciplinary project developed by Robert M. Ellis over more than 20 years, to develop a consistently pragmatic approach to the justification of human judgement. It follows through the implications of the Buddha's Middle Way, rejecting absolute beliefs of a negative as well as a positive type, in the light of the developing modern understandings of uncertainty, scientific method, mindfulness, embodied meaning, neuroscience, cognitive and developmental psychology, systems theory, Jungian archetypes, and democratic political practice.

Diagnosing the central problem of absolutization that interferes with the justification of human judgement, it then seeks to identify the most effective responses to that problem. It does this through the rigorous application of pragmatic philosophy, drawing on a wide variety of evidence. Overall it thus offers a detailed normative ethical philosophy based in the conditions of psychology, and an overall framework to show the relationship of a variety of practices (from mindfulness to critical thinking) to the universal goal of improving each human judgement.

Published

Absolutization: The Source of Dogma, Repression, and Conflict
Robert M. Ellis
(Volume I)

The Five Principles of Middle Way Philosophy: Living Experientially in a World of Uncertainty
Robert M. Ellis
(Volume II)

List of Diagrams and Tables

Diagram 1. The main reinforcing and balancing loops of organisms, as discussed in section 1 — 10

Diagram 2. Robert Kegan's Theory of Stages of Psychological Development, as discussed in section 2 — 72

Diagram 3. Provisional and absolutized cultural features discussed in section 3 — 118

Diagram 4. Some key developments in the practices considered in section 4 — 211

Table 1. Twenty contrasting features of the brain hemispheres (adapted from McGilchrist) — 55

Foreword to the Middle Way Philosophy Series
Iain McGilchrist

The 'Middle Way' Ellis argues for so cogently is far from being a simple compromise between existing polarities, but a departure at right angles to typical thinking in the modern Western world, which looks to me like the path to ancient wisdom.

The perception that objectivity is neither an absolute, nor any the less real for that, is central. Ellis argues for an approach that is incremental and continuously responsive to what is given, rather than abstract and absolute. This is the difference, as he notes, between the pragmatic, provisional, nuanced, never fixed position of the right hemisphere in the face of the absolutism towards which the left hemisphere always tends.

The need for certainty must inevitably lead to illusion, whether in philosophy or in the business of living, and here too Ellis makes clear – as far as I am aware for the first time – the connections between the cognitive distortions known to psychology and the fallacies identified in the process of philosophy.

This is an important, original work, that should get the widest possible hearing.

Dr Iain McGilchrist is the author of The Master and His Emissary *and* The Matter with Things, *and is a former psychiatrist.*
This foreword was originally written for the old Middle Way Philosophy series.

Preface and Acknowledgements

This book is about the cycles of life, beginning with the development of life itself, but then working up to the most complex cyclical processes in human culture. However, it is not just about inevitable cycles to be merely observed, nor doom cycles to be feared (though those do come into it), but also about balancing feedback loops: the ways that we can adjust a maladapted situation through applied awareness, rooted in the general responsiveness of organisms to new conditions.

It's been a fortunate coincidence that during the year or so that I have been engaged in writing this book, I have also become more personally immersed in the wider biological cycles (rather than only the human ones) than ever before. After a morning researching and writing at the computer, I would often spend the afternoon getting my hands dirty, working to turn boggy, compacted Welsh pasture into a forest garden and woodlands with much greater biodiversity and sustainability. There, too, the reinforcing feedback loops of life have had their own momentum: tree guards protecting young trees would be knocked over by deer, and beds mulched for six months to remove the grass would soon get covered in grass again before new plants had had enough time to get established. The difficulties of slowing entrenched momentum were familiar enough, because they are rather like the ones I face regularly on the meditation cushion, whenever I try to change the direction of my mental state. However, the feedback loops of the earth have an immediate tangibility to them that we might miss in more direct reflection on ourselves.

To include a biological dimension to my writing has taken me beyond my comfort zone in the social sciences and humanities, but one that was clearly needed somewhere in this multidisciplinary series, as it is a dimension of Middle Way Philosophy. The results of developing that dimension more fully for me have been stimulating, grounding, and connecting. The historical and psychological dimensions that actually form the majority of this book (after the first section) have been more familiar. Regardless of the area, I take full responsibility for any errors I may have made, which readers are welcome to bring to my attention, whilst at the same time I make no apologies at all for daring to write synthetically and crossing the boundaries of disciplines, which is the basis of my whole approach to practical philosophy.

Robert M. Ellis

Acknowledgements

The author would like to thank the following people for their help in providing feedback on the manuscript of this book: Rodrigo Caceres Riquelme, Nina Davies, Peter Daszak, Dahlia Gibson, Barry Daniel, and Viryanaya Ellis.

Introduction

Every stone in the ground, every person in the street, has a history. That history adds an illuminating dimension to every object or person we encounter. Understanding that history helps us understand the patterns taken by conditions, whether these are the effects of a geology on the soil, the neural patterns of meaning built up in an individual's brain and body, or the connections between concepts and symbols in cultural relationships across the human world. The past tells us, very often, how complex conditions today have built up from simpler ones further back in time, and thus helps us to address that complexity.

However, history also has its drawbacks. Absolutists often repress others, or even wage war, because of their certainty about supposed historical events that they appeal to in support of their beliefs or actions. Such-and-such words were revealed to Muhammad, or Marx said such-and-such, so we now have absolute values that can be applied in the present, regardless of their clashes with other values or their effects on others. Linear historical narratives readily support linear deductive reasoning, in which an authoritative event in the past gives assumed certainty to supposedly justify a brutal shortcut in the present. Such linear history can be readily used to help maintain relationships of power, for instance when an ethnic group tries to eliminate its neighbours in reprisal for events 50 years before (as happened in Bosnia during the 1990s), or when a religious tradition claims that the words of its scriptures are 'true', repressing criticism, because they were delivered in the past by the founder.

So far in my writing about Middle Way Philosophy, then, I have been very careful to avoid focusing too much on history. I wanted to clearly differentiate Middle Way Philosophy from religious or other traditions (including Buddhism) that try to justify their claims through appeals to history. I definitely didn't want anyone to think that Middle Way Philosophy was justified because of what any particular authoritative individual said at any point in the past (including myself), or because of any particular revelatory event. In science, people have largely made that distinction: they don't think the theory of gravity is correct *because of* Newton. Gravity happens anyway, however illuminating or otherwise Newton's account of it may be. However, in many other spheres, particularly religion, they often have not made that distinction. That has made it all the more important to emphasize that Middle Way Philosophy is a theory justified provisionally by its compatibility with practice, and then in the longer term by our very experience of practising it. The Middle Way happens anyway, regardless of the history of talking about it. This point seems especially to be missed by both academics and religious practitioners of the kind who get bogged down in mere scholarship within the restrictive walls of a given tradition, heatedly disputing who really said what, as a shortcut substitute for considering the actual helpfulness or otherwise of what has been said.

However, I have also continued to find history illuminating and fascinating, and to be drawn to writing about the historical aspects of Middle Way Philosophy. Eventually it became clear to me that systems thinking offers a way of doing this whilst avoiding the linear traps that so easily accompany the philosophical or religious use of history. I thus decided to write *systemic history* rather than linear history. Systems are wider sets of relationships between all the people and things we interact with, engaged with at different levels, whilst linearity focuses only on particular objects and their track through time, constantly reinforcing a single framing set of assumptions about them.

Systemic history is the kind of history that constantly strives to put our accounts of specific events in a wider context of relationships and perspectives. Systemic history does not seem to have been written before – at least not by anyone using that term, although one can find some approaches to it in some recent 'big' world histories, or in histories that take unexpected new angles on the complexity of the past, such as tracking the story of a commodity. Systemic history has several identifiable features: it is multidisciplinary, deals with cycles of events as well as with unique sequences of them, deals with nested systems at different levels, and uses narrative about the past to stimulate new reflection on complexity.

Systemic history is multidisciplinary, because it is not concerned only with a restricted set of kinds of causal processes. The typical concerns of much traditional history are political or socio-economic, occasionally stretching to philosophical, religious, and artistic factors. Some more recent history has engaged with the role of science: for instance, the effect of human biology on the human story, or the effects of past climate changes. Only a few pioneering writers have recently started to include neuroscientific perspectives in their history. The more of such differing perspectives can be incorporated into a history, the closer it gets to systemic history. In this book I will be particularly drawing on biology (specifically systems biology) and developmental psychology in addition to the mixture of disciplines that has informed my previous books in this series: philosophy, religious studies, psychology, the linguistics of embodiment, and the systems perspective.

Systemic history deals with cycles of events rather than only with unique sequences of events: for instance, the typical cycle of a human life-span, constantly renewed by new births, or the general features of evolutionary development replicated in the development of new species. Of course, these cycles are a matter of generalized description of a kind that is often considered the preserve of science rather than history, but they are also history. Every past human life-span, for instance, is historical, and needs to take its place within the larger picture of history, as well as within that of science, both in its specific and its generic features. In this book, I will begin with the biological features of organic life, moving then into the psychology and neuroscience of individual humans, and then to the development of human culture. That will require a consideration both of cycles within broader history, and of the more general sequence of human history as a whole.

Systemic history deals with nested systems: that is, with larger and smaller systems where the smaller systems form a part of the larger ones. For instance, the meditation practice of an individual human forms one kind of system, that can be placed within the system of that individual human. This individual human is in turn part of human society at a variety of scales, and of ecosystems in which she participates. The universe as a whole is the largest-scale system we can contemplate, whilst sub-atomic particles offer the smallest. In this book, however, I will be largely concerned with the human mind-body system and its relationships with the most immediately adjacent system levels. To talk about nested systems it is impossible to adopt a simple chronological narrative. Instead, one needs to constantly tack about like a yacht sailing into the wind, talking about the history of one system followed by another on the same level, then perhaps switching levels.

It is this roving about between different system levels that makes systemic history a way of communicating complexity. One of its main aims is to disrupt any tendency the reader may develop to assume that one way of describing things in one context is the whole story. By moving on to another factor at another level, we should be able to more easily see each element of a larger, more complex system in relation to the others.

The systemic history in this book is that of something that could not really be historically explained other than systemically – the Middle Way. As I have introduced it in the two initial volumes of this series, the Middle Way is a response to absolutization, navigating between any two opposing absolute extremes in order to find the best point for learning from experience. The two extremes to be avoided can be found as twin gatekeepers on either side of any judgement whatsoever, each claiming to have the whole story whilst the opposite is denied. To find the way between them, however, we need the five principles as explained in volume II of this series: scepticism, provisionality, incrementality, agnosticism, and integration. These five principles are all themselves aspects of the greater contextualization required in systems thinking: taking into account uncertainty, considering a wider range of possibilities, applying incrementality as a property of systems, recognizing the limitations of rigidified linear perspectives, and avoiding the 'sub-optimization' of sub-systems (their being less well-adjusted than they might be) in conflict with a wider system.

It is thus not surprising if the development of conditions over time that makes all these practices possible only becomes clear when one considers the relations between systems from multiple perspectives – rather than being restricted to the merely linear perspectives of, say, Buddhist scholarship, or evolutionary biology. If we could explain the Middle Way satisfactorily only in terms of the textual and traditional authorities of one religious tradition, or only in terms of the genetic inheritance of species winnowed by trait-selecting environments, we would be missing the Middle Way and its point, by focusing only on one side of the entrenched conflicts of belief that the basic motive of the Middle Way is to resolve.

To discuss the systemic history of the Middle Way, then, this book sets out to follow the development of several kinds of interrelated conditions. There are four of these: the roots of the Middle Way in organic systems, the development of the Middle Way in stages of human psychology, the cyclic waves of absolutization and Middle Way responses in human culture, and the development of practices that can enable Middle Way judgement to actually occur. In brief, one can think of these as biology, psychology, culture, and practice. One cannot have any of these without the others: biology provides the basic bodily conditions, psychology the development of neural connections in the human brain, culture the continuing conditions that sculpt those neural connections through experience, and practice the individual motives to continue developing and using those connections.

The first, biological, section of this book traces the basic conditions for the Middle Way in the reinforcing feedback loops that make living organisms continue, together with the balancing feedback loops that help them adapt to their environment. One can find the roots of these two kinds of feedback loops in organisms from the very beginning of life (and perhaps even before it in some respects). As we follow the development of increasing complexity up the tree of life, however, those feedback loops are constantly echoed. The more organisms turn to reinforcing feedback loops to try to sustain themselves, the more they run into conflicts with their environment, which in more complex organisms also become reflected in internal conflicts. To deal with those conflicts, they develop balancing feedback loops, only for more conflicts to arise as conditions change or competition moves in. In more complex organisms, these two types of feedback loops become the dominant functions of the two hemispheres of the brain – the left reinforcing and the right adaptive. This enables the same feedback patterns to recur in human psychology, history, and culture.

As we will see, this pattern of feedback loops is shared both by the development of species at what biologists call the phylogenetic level (i.e. the level of species), and the development of individuals at what biologists confusingly call the ontogenetic level (which simply means the level of individual organisms, and has nothing to do with their 'ontology', as the term is used more philosophically). Biological evolution is obviously part of this recurrent, cyclic pattern in systemic history, but not the whole of it. Rather than telling a merely evolutionary story, I want to put evolution in a wider framework of significance: one in which human actions and judgements are driven not only by imperatives of survival and reproduction, but also by more complex needs. Moreover, that wider framework is one where the drawbacks of assuming biological determinism of human actions become ever clearer, because of the emergent properties that arise with ever greater levels of complexity. Nor should that framework be supplanted by one of metaphysical freewill, which would be merely a reactive swing of the pendulum – but rather we need to recognize and accept the ambiguities created by a systemic biology as opposed to a simplistic mechanistic biology, and follow through a rigorous agnosticism between both freewill and determinism.

My second, psychological, section follows the continuing impact of reinforcing and balancing feedback loops in the development of each individual human being in response to their culture and environment. The history in this section is thus once more a cyclic history, replayed as each of us grows. The reinforcing feedback process tends to make our development reach a series of plateaus as we reach a mode of thinking and behaving that works sufficiently well, given our bodily development and situation, for the moment. However, these temporary points of stability are disrupted by new conditions that begin to make our old mode frustrating, and trigger adaptive balancing loops. After a period of transition, these balancing loops then settle back into a new phase of relative stability. This punctuated equilibrium in human psychological development was first tracked by Piaget, and has more recently been pursued into adult development by Robert Kegan.

It is by this means that we gradually ratchet up the complexity of our individual responses to our environment. When we settle into a particular stage of development, my thesis is that key judgements shaping our values and our view of ourselves start to become absolutized (absolutization – the assumption that we have the whole story – was the topic of the first volume in this series). A transition between stages, however, requires provisionality over more of these judgements. To reach the most complex level, which Kegan called the interindividual stage, we need to adopt provisional approaches to all the major judgements that shape our values and self-view. This is very far from reaching a perfect view of 'knowledge' – indeed, it is almost the opposite, that is, a substantial acceptance of the continued imperfection of our judgements. The application of the Middle Way is not restricted to this stage of greater complexity, and indeed can be done at any point, but it is applied more comprehensively and consistently by this final stage.

This survey of psychological development will show more fully how the biological roots of the Middle Way reach their fruition in individual judgement. However, that individual development is also constantly in response to the culture in which we live. Human cultures can encourage or arrest the development of individuals, offer or withhold helpful resources, and socially reinforce or block change. The culture in which we live in the twenty-first century is itself the result of a tremendously complex process of cultural development, which has also followed the pattern of reinforcing and balancing feedback loops. Our own individual level may often lag behind that of the leading edge of our culture, but it is very much harder to be in advance of it.

My third section, then, attempts to track this development of human cultural complexity. It does so by identifying a variety of sub-systems that have arisen at various points within human cultures, each of which has offered new complexity and new adaptive potential, but each of which was then followed by a rigidification process as increasingly absolute judgements were applied to it. When this rigidification (or reinforcing feedback loops) became too maladaptive in the face of new conditions, though, frustration has resulted, and new innovators have managed to gain support in adapting the culture with yet another complex sub-system. The pattern is then repeated. For instance, the development of religious archetypes early in human development offered important developmental potential by allowing us to maintain inspiration over time, thus fulfilling plans, being open to new ideas, identifying long-term threats, and feeding helpful relationships. However, this balancing feedback loop (as I tracked in my recent book *Archetypes in Religion and Beyond*) soon became rigidified by the projection of these archetypal symbols into objects of belief: archetypal gods who helped us recall more sustainable qualities of mind, for instance, became supernatural agencies who intervened to help us pass our exams.

This third section may at first sight look a bit more like 'history' as we know it than much of the rest of the book, beginning with questions about early human development in the paleolithic and neolithic, and working onwards from there. However, once again the historical narrative will be very far from a straightforward chronological one. Rather, what I want to track is a series of nested and interrelated processes, with different decisive events but considerable overlap in their periods of development. To some extent I will have to falsely separate out the religious, scientific, technological, and political strands of human culture, for instance, but this is obviously not because one of them stops before the other starts. At all points we need to try to be maximally aware of this interdependence and complexity, though obviously not to an extent that it becomes overwhelming – there is a Middle Way to be followed here too.

It is my fourth section, however, that focuses on the history of the Middle Way as *practice*. Here we enter the ambiguous zone of human responsibility, no longer talking about feedback loops that occurred in cultures as a whole, but rather about the ways that some individuals or groups have to some extent recognized the drawbacks of reinforcing feedback loops, and tried to create the conditions for balancing ones instead. If my history in the third section is a rather depressing one of balancing processes constantly being appropriated by more absolutization, a focus on practice can offer more hope.

However, my emphasis in this book is not on the effects of practice (as it was in *The Five Principles of Middle Way Philosophy*), but rather on the *history* of practices. It is one thing to consider how meditation or democracy can take their place within an overall scheme of practice, but a somewhat different one to ask how meditation or democracy themselves have developed over time. This development takes place within human culture, so I will be effectively considering the history of various cultural subsystems, all of which can be placed, not only in the system of 'practice', but also in relation to many other systems. These systems may be seen biologically, psychologically, socially, culturally, philosophically, religiously, and probably in other ways too. For instance, meditation has primarily developed in the context of religious traditions, but is now increasingly being investigated and used in relation to its biological and psychological effects. Democracy may be understood primarily as a political system, but one of its key conditions for success is biological – a sufficient limitation on human stress levels so that they do not revert to shortcut voting choices.

One of my key emphases in *The Five Principles* was the ways in which practices that people usually do not consider in relation to each other (like meditation and critical thinking) are interdependent and support each other's effects. To consider the history of these practices should provide a further perspective on that interrelationship, because it will allow us to become aware of it not only at one specific time, but over time. For example, we can consider not just how ethical observance might help us prepare better for mindfulness or bodywork, but how past practitioners in the Hindu and Buddhist traditions have used it in exactly that way. We can also see how past philosophical enquiry, together with the developments of education and democracy, have created the conditions for critical thinking practice as such – which extends some of the skills long used by philosophers beyond the relatively narrow and highly abstract contexts in which they were previously used.

Of course, in considering the history of practices, we will also need to consider some of the ways they have been appropriated by absolutizations: meditation is idealized, the arts are sentimentalized and commercialized, democracy decays into anocracy, and critical thinking is deployed selectively with taboo areas where thinking is forbidden. This is an unavoidable aspect of the history of these practices. However, at the time of writing, all these practices are still available to us. Considering their history has the largely practical goal of allowing us to experience and participate in a positive continuity. Every successful past practitioner is a potential source of heroic inspiration for our own practice, as well as a source of information on more or less successful approaches to it in various conditions.

By the end of this book, then, I hope to have travelled, at least from the roots to the crown of the Middle Way, even if I cannot exactly claim to have gone from the beginning to the end of it. Although I will have traced aspects of the path for many people, the path as a whole, being systemic, begins not in the past but in your present experience. It is not the Buddha's enlightenment or the first scientific discovery that begins the Middle Way, but your next judgement. Nor is there an identifiable end to the Middle Way that does not consist of abstract metaphysical speculation: there are just intermediate goals receding into the future, not just for you or me but for us. The history of the Middle Way does not end with an Apocalypse or a Revelation, but with openness towards the future.

1. Conflict and Integration in Organic Systems

The main reinforcing and balancing loops of organisms, as discussed in section 1

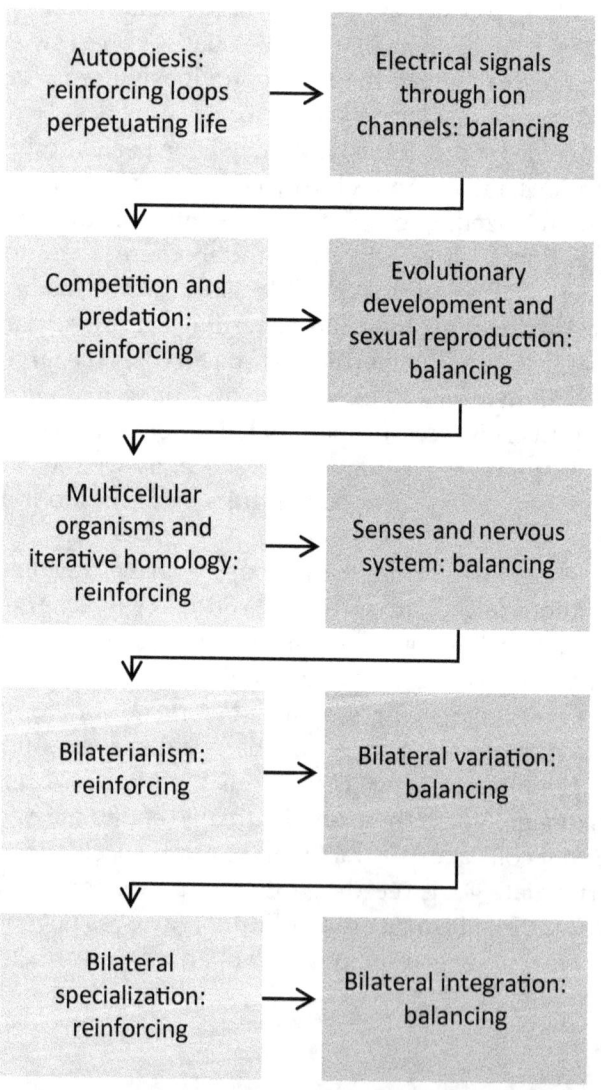

1.a. The Emergence of Self-organization

> *Summary*
>
> The emergence of self-organizing life is the first reinforcing feedback loop, marking the beginning of the succession of reinforcing and balancing feedback loops that are the focus of this book. These feedback loops give the basic prior conditions for absolutization and the Middle Way. Life seems to have emerged in continuity with non-organic matter, through the catalysed differentiation of electrical charges in molecules that results in a self-organized 'inside' becoming distinct from the 'outside'.

The beginning of the history of the Middle Way lies in the origins of life itself, in the form of single-celled organisms emerging about 4 billion years ago. It then continues with a succession of stages in the early development of organic life.

The root conditions for the Middle Way lie in the interaction of two kinds of feedback loop: reinforcing and balancing. In a reinforcing feedback loop, a system responds to new events in a way that defends and reinforces its current form: for instance, any lifeform that preserves itself. In a balancing feedback loop, a system responds to new events in a way that modifies its current form to adapt to the new conditions: for instance, a dog that is trained to beg for its food, whose potential for this new behaviour is then inscribed in its neural pathways.

These two kinds of feedback loop will provide a constant motif for this book at every level. They come not just from biology, but also from systems theory, where they have been observed in many different contexts. As with all the empirical claims in this book, though, there being such a distinction is a provisional belief based on its immense explanatory usefulness so far. There are no ultimate claims intended about whether feedback loops, or any other biological phenomena, 'really exist'; only that a theory that takes them into account is of evident explanatory and practical value. *Phenomena* are how things appear, and this is what the sciences deal in – not *noumena*, which is how they ultimately are.

Although they only start to appear at a greater level of complexity than I am beginning with here, absolutization and the Middle Way, the main themes of this entire series, also consist of feedback loops. Absolutization is one kind of reinforcing feedback loop, and the Middle Way is one kind of balancing feedback loop. To trace the history of how absolutization and the Middle Way have developed, however, we need to begin with reinforcing and balancing feedback loops at a much simpler level.

The concept of self-organization or *autopoiesis* as a defining feature of biological life was first put forward by Maturana and Varela,[1] and is a basic element of biological systems theory. Living cells are differentiated from their non-organic environments by continuing self-organizational activity that maintains the organism. This can take the form of seeking and absorbing resources (nutrients) or defence against threats to that self-organization. The resulting living system is thus operationally bounded but thermodynamically open. It takes in and gives out energy, but nevertheless maintains a continuing form that is not determined by the environment.

It is important to try to understand the emergence of life in a continuous rather than a discontinuous way: we do not need shortcut explanations such as that life instantaneously began by a supernatural spark. We can understand life's continuity with the non-living environment by considering the ways that even non-organic molecules can exhibit some features of self-organization. These features begin with electrical charge, which is a property of the protons (positive) and electrons (negative) that go to make up the structures of atoms and molecules. Opposite charges attract whilst similar ones repel. Ions, which are charged molecules, play a big part in the basic structuring of organisms.

The phenomenon of oil clumping in water offers a basic non-organic analogy for the formation of living systems under the influence of electrical charge. Oil molecules are examples of 'surfactants' which have both hydrophilic heads and hydrophobic tails – one end has the same charge as the surrounding water, so is attracted to its environment, whilst the other end has the opposite charge and is repelled. When oil droplets in water are agitated, then, the hydrophobic tails will tend to cluster together in a globule, facing inwards, with the hydrophilic head facing outwards, like a circle of wagons defending the group against the outside element.[2]

The oil molecules have, it seems, spontaneously organized themselves so as to maintain a separate system from their environment.

1   Maturana & Varela (1980; 1998).
2   Capra & Luisi (2014) pp. 145ff.

All that has been required for them to do this is not a magical ingredient that discontinuously creates 'life', but rather a particular type of complexity which leads different parts of them to behave in different ways. This kind of complexity is described in systems theory as 'emergent' from the simpler level from which it developed, which is another way of focusing on its continuity.

We do not have to accept the unnecessarily ontological claim of holists that 'the whole is always greater than the sum of its parts'[3] to understand emergence as continuity. We do not know *whether or not* the whole *is* always greater than the sum of its parts, so we should not assume either that the whole is necessarily reducible to its parts or that it is not. In the case of life's emergence, we cannot ever have a complete explanation of life in terms of molecular behaviour, for the simple reason that we can never have complete explanations of anything.[4] However, we can have *adequate* explanations in the sense of ones that allow us to think helpfully in terms of the continuity of processes, and thus respond appropriately to life as though it was interdependent with its environment rather than falsely separated.

As we will see later, the differentiation in surfactant molecules is reflected at a much more complex level in the bilateral form assumed by many multicellular animals, which first developed in symmetrical forms and then differentiated those forms. In the brains of vertebrates, this is particularly reflected by one hemisphere defensively separating itself from the surrounding environment, and the other being attracted to it. In the case of brains this is due not just to a simple differentiation of electrical charges, but to a massively complex system in which the billions of cells that make up the brain are stimulated by levels of charge to allow or prevent further charge, through what Peter Godfrey-Smith calls 'voltage-gated ion channels'.[5] The resulting brain as a whole is capable of moving between both modes as one or the other hemisphere becomes dominant, in a way that resembles the molecular structure of the cells of which it is composed. Either an engagement with the environment allows continual adjustment to it (balancing feedback loops) or an avoidance of the environment allows the organism to continue operating unchanged (reinforcing feedback loops).

3   E.g. discussed in Hoverstadt (2022) pp. 13ff.
4   II.1. (See 'The Old and New Middle Way Philosophy Series' listed before the bibliography in this book.)
5   Godfrey-Smith (2020) p. 31.

In both the cases of the oil molecules and of the lateralized brains, this creates the temptation to talk discontinuously and to slip into absolute boundaries: 'life' versus 'dead matter' where simpler systems become self-organizing, and 'mind' versus 'matter' where more complex ones gain interiority. However, in both cases this discontinuity is unnecessarily imposed on a continuous spectrum between reinforcing and balancing tendencies. The defensive functions have to constantly reinforce their position, whilst the adjusting functions need to be constantly responsive. Each of these functions remains constantly interdependent with the other.

The oil molecules reflect the same principle of organization as that found in biological systems, in nucleic acids and protein folding. In both these cases an external catalyst causes crucial structures to form through the clustering of differentiated molecules.[6] The cell could not form independently of the catalyst, nor could the catalyst have that effect without the presence of appropriate differentiated molecules.

In order to reproduce itself, however, the cell requires not just catalysis dependent on external conditions, like that of the oil molecules, but the capability of autocatalysis. Autocatalysis is created by the acceleration of a process of self-organization, after it has been started more slowly by an external stimulus.[7] After being given an initial push, we continue by ourselves, as when we learn to ride a bicycle as a child, then gradually get faster as momentum builds. This acceleration of the process of self-organization is due to the action of reinforcing feedback loops. In the case of oil droplets in water, the more the droplets self-organize, the more surface is available for floating oil droplets to adhere to, and thus the more the droplets continue to self-organize (until all the oil droplets in the vicinity are appropriated). Again, this basic pattern of acceleration parallels similar ones that can happen at a more complex level: for instance, the more people in a group subscribe to a particular absolute view, the more pressure there is for others to adopt it, until everyone is agreed. A similar pattern might be offered by an exploding population of mice: the mice keep reproducing faster and faster until all the food sources in the vicinity are used up, with the capability of the mice to reproduce increasing the more mice are

---

6 Capra & Luisi (2014) pp. 149–50.
7 Ibid. pp. 150–2.

born. Autocatalysis appears to be basic to the very ability of living cells to develop and reproduce: the more the process continues, the more surface there is to stick to.

Individual cells are autopoietic systems, but these then form further multicellular autopoietic systems (see 1.e), which in turn constantly interact with other autopoietic systems. Life thus depends on a basic reinforcing feedback loop: as bounded systems we take new inputs and incorporate them to help maintain an existing structure, both through metabolic reactions (e.g. taking in food, adjusting body temperature) and the production of molecular components (e.g. new cells to repair a wound).

Much detail remains to be added to this picture. I will talk later, for instance, about the distinction between species adaptation and individual adaptation (although I think there is a common pattern, I am not confusing or conflating the two). The relationship of this story to evolution will also require a lot more discussion. However, the immediate next stage of this systemic history is not yet to discuss these issues, but to consider further important features of self-organized life – namely responsiveness and conflict.

1.b. Electrical Responsiveness and Optionality

> *Summary*
>
> The first balancing feedback loops in organisms were needed for them to adapt to new conditions. This was done through ion channels across their boundaries that later developed into senses. The organism also needed to be able to respond to new information through optionality in the form of meaning (a range of possible states) and provisional belief (states determining behaviour, but that could still be modified). We should not reduce this optionality dogmatically into either freewill or determinism, despite the vagueness and ambiguity of its origins.

So far, we have seen how the development of reinforcing feedback loops in organisms leads to their continuation. However, any organism that relied solely on reinforcing feedback to maintain its self-organization would very quickly perish as conditions change. The adaptation of systems through balancing feedback loops is just as much a basic part of organic life from the beginning as is its continuation.

One key feature of adaptation is that it is gradual, being dependent on a balancing feedback loop rather than an instantaneous change. A balancing feedback loop adds new information into existing patterns of continuing feedback so as to modify them (whether slowly or quickly), rather than creating entirely new effects due to new information. For instance, an earthworm coming to the surface in response to water does not suddenly become an entirely new earthworm because it has modified its behaviour. This is not an obscure or controversial point, just one often neglected in application: instantaneous change does not happen in our experience of systems in the world, even if changes are sometimes very fast.[1]

This point is far-reaching, being the basis of the principle of incrementality, which I discussed in detail in *The Five Principles of Middle Way Philosophy*.[2] The evident fact that change can only occur through the gradual addition of new stimulus means that that there are no toggle-switches or dualities in the *phenomenal* world (as opposed to in our concepts) – only increments. Incrementality needs to be a basic part of our way of thinking because it is a basic

---

1   Relatively fast changes in systems are known as *tipping points*, but these are not instantaneous: see II.3.b.
2   II.3.

feature of systems, and discontinuities in our thinking lead us away from adequacy to systems.

However, balancing feedback cannot occur without the stimulus of new information to the organism that leads it to modify its processes in some way. That requires not only that new information will be available, but that the organism will be able to recognize it and be sensitive to it. At the level of single-celled organisms, this can occur because the membrane that surrounds the cell is not impermeable. Rather the membrane allows the selective passage of ions from beyond the cell, and these ions can then modify the state of the cell. This can be done directly, or indirectly – by modifying the cell's sensitivity to further ions by opening or closing the channel. As Peter Godfrey-Smith explains,

> Ion channels are shared, with variations, across all kinds of cellular life, including bacteria. The reasons for bacteria to build elaborate ports and passageways for ions is not entirely clear. Channels may have arisen initially just to enable cells to adjust their overall charge in relation to the outside – tuning as well as taming their charge. Whenever there is traffic across a system's boundaries, though, it tends to take on further roles. A flow of ions can function as a minimal form of sensing, for example: suppose contact with a particular external chemical opens a channel and lets in ions. Those charged particles can set new events in the cell in motion.[3]

In more complex organisms, these ion channels can obviously develop into senses as we understand them. For instance, the human eye can only operate because the cells of the retina are light-sensitive, which is a development of the capacity of all cells to be sensitive in one way or another. The incremental process by which cellular sensitivity can give rise to a complex sense organ such as the eye has been memorably demonstrated by Richard Dawkins, who has used computer models to show all the intermediate stages from one to the other.[4] I will be returning in 1.f to the development of senses and the nervous system as a channel for electrical responsiveness in multicellular organisms.

Electrical stimulus seems to be at the basis of what Maturana and Varela called 'structural coupling', which refers to any way in which an organism interacts with its environment to allow mutual modification. Structural coupling is thus essential for the organism

3   Godfrey-Smith (2020) pp. 30–1.
4   Dawkins (1996) ch. 5.

to adapt to changes in its environment, and also for it to change that environment. In more complex organisms, structural coupling includes modification of a neural network that one could start to call a 'representation' of the environment. Crucially, that 'representation' is not a copy of the environment, nor is it determined by its environment. Rather, it is influenced both by the environment and by the previous structure of the organism.

Here, Maturana and Varela talk, I think wrongly, of the modification of an organism in response to structural coupling as 'knowledge' or 'cognition', terms that maintain the absolutization implicit in representationalism,[5] and that cannot be incrementalized without equivocation – we either have knowledge, as generally understood, or we don't.[6] Instead of 'cognition' or 'knowledge', I find it far more helpful to talk about meaning and belief as the modification of an organism's structure due to interaction with its environment. Both of these terms offer clear potentiality for provisionality, rather than the danger of making absolute assumptions (of a kind that are incompatible with systems theory) that arises when discontinuous concepts like 'knowledge' or 'determinism' are employed.

Although Maturana and Varela don't use either 'meaning' or 'belief', they do implicitly offer the basis of the distinction between them. When organisms become complex enough, they say,

> *The nervous system participates in cognitive phenomena in two complementary ways.... The first... is through expanding the realm of possible states of the organism that arises from the great diversity of sensori-motor patterns which the nervous system allows for.... The second is through opening new dimensions of structural coupling for the organism, by making possible in the organism the many different internal states with the different interactions in which the organism is involved.*[7]

In other words, organisms develop a capacity for meaning when they change their own internal structure in some way to allow for a greater range of possible states or behaviours in future. This does not mean that they will necessarily ever use any of these possible states or behaviours. In humans, we tend to think of this as the realm of the imagination. Organisms develop beliefs when their structure is changed in such a way that their behaviour in relation to their

---

5   I.3.a.
6   II.1.a.
7   Maturana & Varela (1998) p. 175.

environment develops a new, modified pattern. At this point, then, a balancing feedback loop has become a reinforcing one, at least for the moment – modification due to new information has become a more settled pattern of response. For humans, that pattern of behaviour can include the uttering of explicit statements of belief (such as 'I believe that Monty was lying'), but much more often, even with humans, our understanding of belief comes from behaviour of other kinds (such as a slightly more cautious approach to Monty).

The possibility of meaning is thus there from the beginning of the story of life in its development through self-organization, even before nervous systems develop (see 1.e) to transmit and apply electrical stimulus in increasingly complex ways. Even for single-celled organisms, to self-organize is to repeat structural patterns so as to maintain oneself, and by repeating certain structural patterns we strengthen the associative links between particular kinds of stimuli and particular sorts of structural changes: for instance, the detection of food helps set up the organism to consume it. To enable electrical stimulation to have a helpful effect, we have to build and maintain channels for them both to stimulate and change us.

We may regard the behaviour of organisms, especially simpler organisms, as highly predictable, and thus be inclined to merge the concept of meaning with that of belief. However, even with the simplest organisms there is no total predictability: we do not know exactly how they will behave, we only probabilize and justify our beliefs about them. The more complex the organism, the harder it is to ignore the complexity of the structural changes that intervene between stimulus and response – especially when we are dealing with the trillions of neural connections in a human brain. Without total predictability (which would have to be based on total representation in our own minds when understanding that predictability) we have no grounds for absolute beliefs such as determinism. Organisms maintain an ambiguity locked into their complexity, because what they find meaningful is to varying extents distinguishable from what they believe.

It is important to distinguish meaning from 'knowledge' to account for provisionality. Meaning is still completely flexible. However, belief, too, can be provisional. In belief at the level of complexity represented by humans, there can be provisionality, because the imagination allows us to explicitly formulate potential or hypothetical beliefs prior to committing ourselves to those beliefs through action. The fact that an organism consistently behaves in a particular way also does not tell us whether or not it is well-adapted when it does so – whether it is 'facing reality' or evading it. Belief is a stronger instance of self-organization in response to structural coupling than is meaning, but it is still not 'knowledge', just a more-or-less justified consistency of response due to a more-or-less habitual structure in the organism.

It is this provisionality, not determinism, that allows switches between the reinforcing feedback loops of self-organization and the balancing feedback loops of adaptation. An organism that has made structural changes to suit one environment at one time does not merely continue with the same changes indefinitely. If it does so, it is likely to perish before very long as the environment changes: the millions of mice die when the grain supply is eaten up. Instead, in response to changing conditions, the self-organization also changes. Such change can only be incremental, because it depends on the very same process of autocatalysis that self-organization itself also depends upon. Living organisms are not Proteus of Greek myth, able to instantaneously switch from one form to another, but they are able to start a new modification of structure and then build on it. A plant that has been growing in one direction towards the light cannot retract what has grown already, but can switch its energies in a new direction when the long-term light source changes. A deer whose habitual path has been blocked by a fence at first tries to leap it, but, frustrated, is then forced to change its habitual path.

From the very beginning, then, self-organization contains both the potential for non-adaptive rigidity and the potential for adaptation. It is this differentiation of two forms of potential, I argue, that provides the source of the Middle Way. Please be clear that I am not suggesting that the Middle Way itself is available to the most basic forms of self-organizing life. At this stage we only have, at best, a kind of proto-absolutization in uncontrolled replication of the same structures, and proto-Middle Way in the adaptation of those structures at any given stage. When exactly absolutization and the Middle Way themselves emerge will require further discussion.

What is important at this stage is that we interpret the evidence for the emergence of life in a way that is itself guided by the Middle Way, so that we do not make absolute assumptions that then interfere with our understanding of how the Middle Way can emerge from it. This particularly means that we need to avoid any assumption that 'nature' is operating in any kind of coherent, rationalized form beyond the phenomena of systems, or on the other hand that living systems cannot develop what looks to us like coherence. For the same reason we need to avoid the assumption of either determinism or freewill: we do not know whether or not the self-organization of life or its adaptation is in any sense 'free'. Rigour is required to avoid this habitual and entrenched false dichotomy from creeping into our understanding of the development of life in ways that will then later make our own experience of provisional judgement incomprehensible.

Living systems, from single-celled organisms through to humans, have the capacity to adapt in response to this new information only because of *optionality*: that is, the presence of different possible responses. Sometimes those different possible responses are behavioural, involving an organism having the capacity to do something in response to a stimulus. At other times these responses may consist of different longer-term responses by an organism: for instance, a tree may respond to light by growing in that direction. Such longer-term responses might potentially be broken down into specific events in the structure of the organism, usually due to electrical charges making molecular changes that then influence its growth, reproduction, digestion, movement, or any other activity. For instance, a tree that grows towards the light can only do so because its direction of growth can be modified in response to that light, through the greater direction of sap towards some growing points rather than others. Such directionality of growth is amongst the tree's *options* – the things it can potentially do. Uprooting itself and then blasting off to the moon, however, could not reasonably be described as amongst the tree's options, even if we can imagine it. Single-celled organisms from the beginning of life also seem to have such optionality to at least a small degree. For example, an amoeba on the edge of a zone that is slightly more nutrient-rich may respond to marginal signs of opportunities for nutrition by moving towards it, or it may fail to respond because of the marginality of the signal.

There are at least two philosophical traps that it is important to avoid in using the concept of optionality – the freewill-determinism dichotomy and an over-reaction to vagueness. The freewill-determinism dichotomy is being applied by anyone who insists either that organisms can have no 'options' because the stimuli concerned together with their structure make their responses inevitable, or that such options are limited to humans when they have self-conscious 'choice'. In practice, we have no justification for the absolute claim that any event is inevitable, nor on the other hand that even humans with high levels of brain complexity and awareness have complete control over their choices. This point will be explored in more detail later in this series, in *The Practice of Agnosticism*.[8] The turning of ambiguous information about the agency of organisms into absolutes leads us quickly into absolutizing assumptions that we then apply in speculative beliefs about responsibility, sentience, and so on, when at best we can make probabilizing judgements about the *extent* of another organism's responsibility. That the responses of single-celled organisms are highly predictable does not make them 'mechanical', even if the probabilities of them responding in a particular way are asymptotic – drawing endlessly near to zero or 100% on a graph, but never getting there. As I argued in *The Five Principles*, asymptotes need to be respected as such, but not turned into absolute certainties.[9] We are in no position, in the end, to distinguish between 'determined', 'random', and 'chosen' responses in living organisms: instead we have a spectrum of degrees of predictability and complexity, and need to apply incrementality rigorously in thinking in terms of that spectrum rather than absolutizing shortcuts, however institutionalized these may be.

An over-reaction to vagueness is another philosophical trap that can lead us into absolutizing shortcuts when considering the responses of organisms to stimuli. Options are difficult to determine, and the boundaries of 'potential' in an organism cannot be well-defined. Taking again the earlier example of a tree responding to light, what are the boundaries of how it can potentially respond? Obviously there are some limits, as without some light the tree will die: its options do not include growing in total darkness. An etiolated tree growing weakly towards a limited and distant light-source will also be severely weakened by its conditions. Some options are possible for an organism, but also impose such costs (weakening its structure, for instance) that they would only be adopted *in extremis*. The fact that we cannot always predict *in advance* what options are

---

8 VI.3; iv.4.c.
9 II.2.f.

possible for an organism, though, does not mean that we cannot reasonably talk about options.

Conditionality does clearly play a role in any positive description of what constitutes an option. There needs to be an incremental pathway by which an organism can move from its current condition to any adaptive condition. That incremental pathway follows the processes recognized by biology: namely metabolism, growth, movement, and reproduction of various types. Whether an organism is capable of a particular option thus also depends on the time it has to fulfil it: an acorn has the option of turning into an oak tree only via a series of stages that are each dependent on the surrounding conditions. An option can thus be defined as *a potential pathway of gradually changing contiguous states along which an organism can operate its biological functions, subject to the time and conditions required at each point on the pathway*.

Such a definition covers optionality in the whole range of organisms from single-celled microbes through to humans, with a range of complexity applied to the biological function involved. That range of complexity goes from mere continuing survival or reproduction through to a complex human need such as intellectual integrity: for however complex the need, it will need to be fulfilled by activities of the organism along a pathway of contiguous states.

Such a definition also leaves open whether 'an organism' is being considered at an individual or a species level: the species level being effectively a way of referring to an accretion of genetic changes in individuals that are then winnowed by selective conditions. Evolution provides one way of talking about options, because genetic changes in the structure of organisms are gradual and contiguous changes in state that respond to conditions (even if they respond slowly and indirectly). I will go into this point in more detail in 1.d. However, to talk of evolutionary change at species level is a way of generalizing about many contributory changes at individual level – which of course are also affected by species-level change. These levels of change are completely interdependent, so we need to avoid any dogmatic assumptions about the intrinsic priority of one or the other.

In sum, then, organisms *adapt* through the use of balancing feedback, because they have structurally developed channels by which they can be sensitive to external information about conditions. That sensitivity may involve immediate changes in behaviour, or may alternatively consist of longer-term changes in structure that only occur through indirect selection. Such changes can only potentially follow the pathways of optionality, which are incremental routes to fulfilling biological functions.

1.c. Competition and Predation

> *Summary*
>
> Competition and predation between organisms is in practice unavoidable, and is the source of systemic conflict. Systemic conflict needs to be distinguished from absolutized conflict, which rigidifies conflicting needs into maladaptive repression. Systemic conflict is compatible with homeostatic balance in an organism's response to its environment (which is a kind of proto-Middle Way), whereas absolutized conflict is not.

Given that the emergence of life demands reinforcing feedback in living systems, some degree of predation between them seems inevitable. Reinforcing feedback maintains a living system in its existing structure. To do this, in accordance with the laws of thermodynamics, requires it to take in whatever energy it uses for growth, movement, or temperature regulation. Some organisms gain energy directly from the sun, and a few from geothermal sources, but others do so by ingesting other organisms and turning their material into energy. This creates predation, as well as competition with other predators.

Predation became possible in the early development of single-celled organisms through the development of eukaryotes, around 1 billion years ago.[1] These have a cytoskeleton and an ability to process energy using mitochondria. These developments enabled single-celled organisms to change from being inert chemical processors to having the capacity to actively seek out food. Movement is a necessary condition for predation, since you cannot eat anything unless you can find it first. Movement requires energy processing to fuel it, and a structure that can withstand the stresses it creates.

The development of predation obviously creates a systemic conflict between individuals through a clash of motives: organisms seek to ingest other organisms, but also to avoid being ingested themselves. An accelerating 'arms race' ensues, both at the level of behaviour of individual organisms in relation to each other, and more broadly at the level of evolving morphology and anatomy, as types of organisms compete with each other. Even if organisms are not directly predating each other, competition can also emerge for any resource in limited supply that is needed by the organisms: not

---

1   Keeling et al. (2005).

only food but possibly territory, sunlight, water, minerals, housing materials, or mates.

As with any generalization, one cannot conclude that systemic conflict between organisms is *necessarily* inevitable. One could imagine a situation in which organisms depended only on freely available resources such as sunlight, and where their populations were stable enough not to have to compete with each other. In his novel *The Possibility of an Island*, Michel Houellebecq imagines such a scenario for humans who have genetically modified themselves in the future, so that they require only minerals, water and sunlight.[2] However, plants, who have similar needs, still compete with each other constantly for stretches of soil in which these resources are available.

In practice, however, conflict is unavoidable in the actual organisms that we observe. The ability to compete in one respect or another follows from the basic phenomenon of autocatalysis noted in the previous chapter, resulting in accelerating exploitation of a resource that then excludes (or in some cases directly destroys) other organisms. We only require the conditions of catalysis (a kick-starting outside stimulus) to occur for accelerating autocatalysis (self-motivation) to come into operation, in which those conditions keep replicating themselves. The more this process accelerates, the more other organisms are likely to be predated or out-competed.

The temptation may then be to assume that this systemic conflict between organisms has necessary 'winners' and 'losers', whether at the individual or species level. That assumption can be supported by the principle of *competitive exclusion* in ecology, which states that a type of organism will occupy an evolutionary niche to the exclusion of other organisms, because it is best adapted to that niche. For instance, in a study of two different types of barnacle best adapted to slightly wetter or drier conditions on the coast, it was found that each could only live in the preferred conditions of the other when the other was completely absent, otherwise it would quickly be out-competed by its slightly better-adapted relative.[3]

However, it is not always the case that an out-competed type of organism vanishes completely from a niche where it has been out-competed, particularly at the level of single-celled organisms. A

---

2   Houellebecq (2005).
3   Connell (1961).

low-level, residual stock of alternative micro-organisms is present, ready to spring back into action when the conditions are favourable. As Hibbing et al. comment:

> *Tremendous microbial diversity has been revealed by new molecular methodologies such as metagenomic sequencing and deep microbial tag sequencing. These approaches and others have begun to reveal that underlying the numerically dominant microbial populations is a highly diverse, low-abundance population (described as the rare biosphere). Members of the rare biosphere that are amplified under favorable conditions to which they are pre-adapted can give rise to discrete, abundant populations.*[4]

The discovery of the rare biosphere is particularly interesting as an indicator of the degree of complexity that underlies the phenomenon of competition, and as an indicator of the limited efficacy of *repression* as a strategy, not just at the human level but in living systems generally. When in a position of conflict, we try to *eliminate* the competition: we want to remove the weeds from the garden, destroy the cockroaches, kill the enemy. However, we are most unlikely to succeed as long as the conditions continue that allowed the competition in the first place. The rare biosphere only operates at the microbial level, but also offers a reminder of the resilience of underlying diversity at more complex levels (for instance, long-dormant plant seeds or animal eggs). Where we dominate an ecological niche, the underlying conditions may still allow reversal after the immediate conditions for that dominance are modified. The microbes offering a potential for disease are still there even when our immune system holds them at bay. The seeds of the 'weeds' are still present in the garden, ready to germinate at the right moment. A childhood trauma, long repressed, can still come back to haunt us. An ethnic group that was hunted down and subjugated by its conquerors may still come back to re-assert rights over its territories many generations later.

Thus *systemic* conflict is always empirically a feature of life, but the success of repression as an attempt to eliminate others in conflict is not. The attempt at repression is the effect of autocatalysis, through which organisms that are achieving dominance get locked into reinforcing feedback loops rather than adjusting them. In more complex organisms, we may then begin to associate those reinforcing feedback loops with repressive motives, as the organism's

---

4   Hibbing et al. (2010), citing Sogin et al. (2006).

access to a variety of possible sensori-motor patterns (the beliefs available to it) becomes limited by those dominant behaviours. For instance, we may continue to waste energy and pump up anxiety swatting phantom mosquitoes whenever our skin tingles slightly. Repressive behaviour may then be either genetically inherited or learned through experience, locking organisms into an endless struggle for life – but one in which no success is guaranteed.

Conflict as a feature of life can thus be differentiated into two types. The most basic and unavoidable type of conflict is just a result of predation and competition between organisms: we can call this *systemic conflict*. However, the other type, what I will call *absolutized conflict*, is the effect of repression in which fixed beliefs lock an organism into limited responses. A lion that kills a zebra for food is in systemic conflict with it, but a lion that does not consider killing a wildebeest instead when zebras are scarce is in absolutized conflict with its own and its pride's needs. For a lion to stop killing zebras and become a vegetarian would require an impracticable degree of change to its brain structure and physiological needs, but for it to break out of a rigid pattern of hunting requires only a small nudge on the ambiguous periphery between 'freewill' and 'determinism'.

It is very difficult to say how far back in the evolution of organic complexity the roots of that distinction lie, but it is possible that we can track it right back to single-celled organisms. Even the simplest organisms require a balance between predation and defence, and between activity and inactivity. An amoeba that hovers on the edge of a nutrient-rich zone without taking advantage of it may starve, but one that spends all its absorbed nutrients searching for further nutrients that are not available may also starve. This balancing, known as *homeostasis*, enables organisms to play a balanced role in an ecosystem that then becomes self-regulating as all the different sub-systems in it adapt to each other. Homeostasis is not the Middle Way – only a starting point for reaching understanding of it – but it does resemble the Middle Way in its relationship to conflict. Systemic conflict is compatible with homeostasis because it is part of the system, and of the regulation of each organism in relation to that wider system. However, absolutized conflict is potentially disruptive of that system.

An amoeba that is merely fixed in its ways will obviously not greatly affect the ecosystem that it is part of: it will merely perish and be replaced by others that are better adapted. However, the amoeba differs from more complex animals such as humans in having a much more limited power to impact its environment. An inflexible lion may temporarily have a bigger impact – but it, too, will probably just be replaced by other more flexible and adapted lions. A human community with inflexible leadership, however, may be totally destroyed by that inflexibility: as seems to have been the case for instance in the eventually extinguished Norse community of Greenland, who seem to have developed a taboo against eating fish when this was the most obvious source of food in a harsh environment.[5] A technologically advanced but inflexible human community, as we know, may take the whole of the existing ecosystem on earth with it when it destroys itself. The further up the scale of complexity we go, the more absolutizing conflict becomes relevant, and the less conflict is merely an unavoidable aspect of the living system.

---

5   Diamond (2005) pp. 229–30.

1.d. Sexual Reproduction and Adaptivity

> *Summary*
>
> Sexual reproduction offers a further method of balancing feedback for organisms that enables them to be adaptive, though at considerable short-term cost. To understand this adaptivity, however, we need to apply scepticism to recognize our full uncertainty about future states. Such a wider understanding of adaptivity also requires that we recognize both its extension beyond fulfilling the goals of survival and reproduction, and the role that this gives to individuals as well as species.

Though poet Philip Larkin famously wrote that 'Sexual intercourse began/ In nineteen sixty-three…/ Between the end of the *Chatterley* ban/ And the Beatles' first LP',[1] he seems to have been out by about 2 billion years. Sex provided the next major stage in the development of life after the initial development of single-celled life. Surprisingly, this development precedes that of multicellular life, and seems to have occurred first amongst eukaryotes – that is, single celled-organisms with a distinct nucleus containing their DNA. This distinct nucleus may have originally developed from the combination of single-celled organisms through parasitism or symbiosis.[2] This development created the possibility of organisms interacting to recombine their DNA, both to repair it after damage, and to create new independent organisms with distinct DNA resulting from a combination of that of their parents.

Bacteria that evolved prior to this development 'reproduce' just by splitting into two. However, the two resulting bacteria are then normally genetically identical, limiting adaptability. Single-celled organisms such as bacteria can compensate for this lack of genetic change at the point of reproduction by being very short-lived and potentially multiplying very rapidly in favourable conditions, so that errors in gene transmission, together with possible exposure to external mutagens, allow them to remain highly adaptable.[3] There are also some limited ways that bacteria can repair their DNA, either by direct gene transfer or by taking up DNA from their

---

1   'Annus Mirabilis': Larkin (1974).
2   Lodé (2012).
3   Watford & Warrington (2022).

environment, but these methods remain incomplete compared to those developed by eukaryotes.[4]

The crucial new element in eukaryotic reproduction is the process of *meiosis*: that is, the recombining of the two complete sets of genes from the parents into one new complete set of genes that determine the form of a new organism. This shuffling of the genes clearly has an adaptive function that becomes important when the environment is rapidly changing. Rapid changes in the environment can produce stress on the organism, for instance through overcrowding, scarcity of food or other resources, or damage to DNA, and such stresses have been shown to stimulate meiosis in eukaryotes.[5]

The widespread use of sexual reproduction amongst organisms has led to debate amongst biologists puzzled by an apparent conflict between its longer-term adaptivity and its shorter-term disadvantages.[6] Sexual reproduction is slower than asexual reproduction, so becomes disadvantageous to a species that is expanding to take advantage of a relatively stable and advantageous evolutionary niche. The random shuffling of genes with each recombination also means that helpful genes can be lost and unhelpful ones passed on, as well as the converse. Sexual reproduction thus increases the losses from maladapted offspring, even if it produces a few that are better adapted to new conditions. Moreover, sexual reproduction takes a lot of resources and energy for an organism: for more complex multicellular animals, this includes seeking out and selecting a mate, whilst species that don't do this but merely release their unfertilized seed into the environment (such as many plants and marine animals) have to do so in enormous quantities to enable chance fertilization to occur.

However, a wider awareness of the impact of reinforcing feedback loops in relation to balancing ones can help to make sense of this problem. Asexual reproduction creates a reinforcing feedback loop in which the same processes are repeated without adaptation. This might have shorter- or even medium-term benefits to a species. For instance, aphids can reproduce rapidly through parthenogenesis (asexual reproduction) during the summer, when food is abundant, but sexually during the autumn. Sometimes there are distinct populations that reproduce asexually for an extended period, but

---

[4] Lodé (2012).
[5] Bernstein & Bernstein (2010).
[6] E.g. Visser & Elena (2007).

the more typical pattern is to mix sexual and asexual forms of reproduction.[7] Either way, it is clear that asexual reproduction is a strategy that can only work effectively in certain limited conditions, just like any other form of reinforcing feedback.

The process of random mutation of genes from a parental starting point, together with natural selection (that is, basic Darwinian evolution) may be a highly blunt instrument for adaptation, but it nevertheless has a basic balancing function, and it is strongly facilitated by the genetic shuffling that sex enables. To understand the adaptive value of sexual reproduction we have to take a long-term or diachronic view, not resting content with beliefs that function only within a limited horizon of circumstances, but applying the practice of scepticism[8] to remain aware of the uncertainties that lie beyond that horizon.

A recognition of the value of scepticism needs to be applied to our very understanding of the idea of *adaptivity*, which will become more important later when we move on to looking at the value of balancing feedback loops for multicellular organisms, particularly including humans. *Adaptivity* is not simply *adaptation*, which is the process by which organisms continue to survive and reproduce in a specific changing environment, but the ability to do so in further possible environments that have not yet been encountered. This concept of adaptivity will need to be applied in a variety of contexts as we go on to consider the increasingly complex developments of organisms. For that reason, the remainder of this chapter involves a short diversion from the roughly chronological systemic history of the evolution of life I am following in this section as a whole, to deal with key questions of adaptivity in general.

Most basically, scepticism implies that we cannot determine the spatial or temporal range within which the idea of adaptivity should be applied. What may appear to be wastage and inefficiency within one range may turn out to have crucial value when that range is extended further. Whilst sexual reproduction provides an initial example within traditional Darwinian terms, human life offers many further and more complex examples of it. Take the neoliberal emphasis on 'efficiency' in public services, that then makes it impossible for health services to cope with crises, or teachers to

---

7   Hardie (2017).
8   II.1.

provide effective education. 'Efficiency' here means functioning without wastage within a restricted set of conditions, but not allowing for others beyond that restricted set. Biologists puzzled by the 'inefficiency' of sex bear some resemblance to utilitarian economists puzzled by education or the arts. To overcome the puzzlement, we do not need appeal to metaphysical sources of value, but we do need recognition of the role of the individual perspective that is capable of becoming aware of a wider range of possible conditions. Even though the value of sexual reproduction can be explained in Darwinian terms, the wider concept of adaptivity needs to be extended in that process beyond the merely Darwinian.

Darwinian evolution has become so strongly established in biology, that many people's first association with the term 'adaptation' will be of the evolutionary adaptation of species rather than of the adaptation of individuals. However, adaptation is a term that can be used of a process of change in the genes of a species, in a population, or in an individual. The Darwinian tradition has insisted that biological fitness can only be judged in terms of the criteria that would directly help to maintain the continuation of a *species* – namely survival and reproduction. In evolutionary terms, then, adaptation is said to take place when organisms maintain or improve their survival and reproduction in a particular evolutionary niche despite changing conditions.

Given that the individual and species levels of adaptation are entirely interdependent, however, there is no wider reason for this restrictive view of the criteria for fitness. I thus propose that we should not be constrained by merely Darwinian views of fitness when considering adaptivity, particularly at the human level. At greater levels of complexity, organisms become fitter in relation to their context whenever their needs in general continue to be met in changing conditions – not just their survival and reproduction. Maslow's hierarchy of needs includes a range of needs at greater levels of complexity than survival and reproduction, including security, social relationships, confidence, and the bundle of more intellectual and moral developments that Maslow called 'self-actualization'. In some cases, a more complex organism might choose to prioritize these more complex needs over those of survival and reproduction, as we find for instance when an artist chooses not to have children, and perhaps even to sacrifice their own health as an individual, to pursue their art. Such an artist would not be maladapted to their environment, but rather supremely adapted in a peculiarly complex way. Our understanding of adaptivity needs to be adequate to the whole range of organic development from the beginning, rather than to exclude the human world.

One objection to this wider view of needs might be that a more complex organism that prioritizes needs beyond survival and reproduction might well be out-competed by another one that does not. For instance, 'preppers', a social group who focus on preparing for survival after the breakdown of civilization, will be at an advantage if civilization does break down, compared for instance to groups of artists who prioritized 'self-actualization'. Of course this might be the case, but the scenario is hypothetical. If we do not know how future conditions are going to be, we do not know what needs to prioritize. Broadly speaking, the further up the pyramid of Maslow's hierarchy are the needs we prioritize, the more vulnerable we are to our ability to fulfil those needs being undermined by the loss of the more basic conditions required to meet the needs lower down the hierarchy. Artists may stereotypically starve for their art, but can only do so to a limited extent before they die, along with their art. However, in an alternative scenario in which the basic conditions of the food supply remain reasonably stable, the artist has fulfilled higher needs that the 'prepper' has not fulfilled. Abandoning the strict Darwinian view of needs is an implication of fully recognizing our uncertainty about the future of complex systems.

This perspective can be reinforced by considering some of the academic discussions that have tried to define and clarify the nature of adaptivity, such as those of Ezequiel di Paolo and of Barandiaran and Moreno.[9] These do not explicitly question the Darwinian constriction in the notion of needs, but they do talk about 'intrinsic teleology' and also suggest that biologists should apply a notion of normativity in relation to organisms (albeit one that is nevertheless still 'naturalized' and thus not properly prescriptive or universal) – that is, that organisms can be better or worse adapted. I prefer to avoid this kind of language, which might contribute to obscuring some key points in Middle Way Philosophy. Organisms clearly have goals, and beliefs which are in concert with those goals, but to call those goals 'intrinsic' or 'a teleology' is likely to obscure the way that those goals and accompanying beliefs may change. The

---

9 Di Paolo (2005); Barandiaran & Moreno (2008).

common elements that frame those goals and beliefs may well involve survival and reproduction, but are not limited to them. Instead, a phenomenological approach should be sufficient: any consistent needs that appear to motivate organic adaptation can be accepted as such, without us needing to speculate on their origins.

Where these academic discussions of adaptivity are more helpful is in identifying exactly how a complex system may be said to adapt. Di Paolo points out that a capacity to maintain mere survival would offer an absolute criterion, unable to be incrementally modified so that an organism is judged to be better off if it becomes more secure in its adaptation. His definition of adaptivity is quite complex, but worth quoting:

> A system's capacity, in some circumstances, to regulate its states and its relation to the environment with the result that, if the states are sufficiently close to the boundary of viability,
>
> 1. Tendencies are distinguished and acted upon depending on whether the states will approach or recede from the boundary and, as a consequence,
>
> 2. Tendencies of the first kind are moved closer to or transformed into tendencies of the second and so future states are prevented from reaching the boundary with an outward velocity.[10]

Much depends here on the idea of 'viability', which is open to a traditionally Darwinian interpretation in terms of the maintenance of survival. A 'viable' organism is one that *can* survive, but this viability is not guaranteed by prioritizing the most evident means of survival. Even if we do see adaptivity initially in indirect Darwinian terms, as soon as we move into an incremental model of this kind (by talking about moving closer or further away from 'the boundary of viability'), a much more complex set of possibilities has to be considered. A relatively well-adapted organism is not simply one that has succeeded in hanging on without dying, nor even one that has managed to pass on its genes. It is one that has expanded the range of possible conditions that it addresses to a greater extent (by moving its tendencies further away from the boundary of viability), requiring it to adopt one or more of a range of strategies that have some probability of aiding its survival in the longer term. Since we do not know what those strategies are in advance, but can only describe them with the benefit of hindsight, we cannot possibly

---

10   Di Paolo (2005).

limit them to those that can only be directly ascribed survival or reproductive value. The more we start to explore the implications of the Darwinian constriction of needs in a more fully diachronic and sceptical context, then, the wider those needs become.

The implication of recognizing a wider range of needs is to include both evolutionary and individual development within the same spectrum of adaptive development. Evolutionary change still depends indirectly on the behaviour of individuals, although it operates directly through genetic mutation and natural selection, just as the starting point for individual development has been created by evolution. This can be acknowledged without any necessity for the now widely criticized claim by Haeckel in the nineteenth century that 'ontogeny recapitulates phylogeny' – that is, that individual development must follow exactly the same course as evolutionary development. It was Walter Garstang back in 1922 who reversed Haeckel's claim, to put the relationship on a clearer footing: 'The phyletic succession of adults is a product of successive ontogenies. Ontogeny does not recapitulate phylogeny – it creates it.'[11] Ontogeny refers to the genetic inheritance of individuals rather than their behaviour, but genes are practically useless unless they are used. Although genetic mutation of an existing genetic inheritance occurs in a random fashion, leading to changes that are usually harmful, the success of any given mutation depends on the ecological niche that it is used in, and this niche may also be modified by the organism. Modified organisms may be at an advantage by staying in an existing niche that has been modified by changes in the surrounding conditions (as mammals were after the event that wiped out the dinosaurs), or they may instead actively move into a new niche that fits their modification (for instance, an animal slightly better adapted to climbing may move into the trees). They may indeed also modify their niche (as humans did through agriculture).[12]

If we are to understand organisms in systemic relationship to their environment, rather than in abstracted isolation from it, we cannot ignore the impact that the behaviour of organisms has even on evolutionary development. Certain crucial transitions in the evolution of animals make this especially clear: there must have been

11 Garstang (1922).
12 Cooper (1996) pp. 73–8.

a first fish that crawled out of the water, or a first mammal that climbed a tree. Even if these processes were incremental (a given animal went up the tree only a little way for a short time), they offer the longer-term basis for decisive tipping points.[13] As Karl Popper put this:

> Every behavioural innovation by the individual organism changes the relation between that organism and its environment: it amounts to the adoption of or even to the creation by the organism of a new ecological niche. But a new ecological niche means a new set of selection pressures, selecting for the chosen niche. Thus the organism, by its actions and preferences, partly selects the selection pressures which will act upon it and its descendants.[14]

The role of activity by organisms in the evolutionary process also reinforces the need to avoid any deterministic dogma in our interpretation of evolution. We do not know to what extent the organism may 'choose' an individual path that impacts the whole future development of its species, and thus need to accept the ambiguity of this position, rather than leaping to the metaphysical assumptions of either freewill or determinism.

This individual role in adaptivity also depends on the sufficient development of nervous systems in organisms, and on the degree of autonomy that increasingly complex nervous systems can develop. Whilst sexual reproduction amongst eukaryotes introduces a type of adaptivity that can be understood overwhelmingly at a phylogenetic (species) rather than an ontogenetic (individual) level, as we move incrementally up a scale of complexity from this to human judgement, adaptivity becomes increasingly dependent on autonomy because it is increasingly individualized. The basis of this autonomy will be discussed further in 1.f, after we have introduced the role of the nervous system.

---

13   See II.3.b on the phenomenon of tipping points in systems.
14   Popper (1992) p. 180.

1.e. Multicellular Organisms and Homology

> *Summary*
>
> Multicellular organisms extend the demands of adaptivity to internal as well as external environments. They are made possible by homology, the reproduction of similar characteristics either within the organism or in making new organisms. This is a form of reinforcing feedback loop. It enables specialization, reproduction, and bilaterianism, all of which can develop into maladaptive patterns, but which also provide the conditions for subsequent balancing feedback.

The development of multicellular organisms, or *metazoa*, offers a crucial further step for organic life-forms and for the complexity of their self-organization. The combining of single-celled organisms into bigger organisms first seems to have occurred between 1500 and 800 million years ago, and to have happened in two different ways: a eukaryotic cell could incorporate other cells within its existing structure, or cells could divide but refuse to separate from each other.[1]

A multicellular life-form, such as a human being, is a city – a bustling community of vast numbers of cells cooperating as part of a wider system. The shared motive for the maintenance of this community is the continuing satisfaction of the needs of each cell, so it seems on the face of it to involve reinforcing feedback. However, to work together, the cells also need sensitivity to each other as made possible by electrical responsiveness, requiring constant balancing feedback within the multicellular organism. At this point, then, the two types of feedback start to apply both internally and externally: we constantly assert our collective metazoic boundaries through reinforcing feedback, just as do our individual cells, and indeed just as do the sub-systems within a body (such as organs). At the same time, that continuing structure can only be maintained at any of these levels through constant adaptation. We adapt to changes both in external conditions beyond the metazoic organism, and internal conditions within it.

The reinforcing feedback that maintains multicellular organisms is the constant reproduction of similar characteristics. This is *homology*, defined by Van Valen as 'a correspondence between two

---

1 Godfrey-Smith (2020) pp. 38–9.

or more characteristics of organisms that is caused by continuity of information'.[2] This could involve any reproduction of structure whatsoever, from the similarity of DNA in the cells of a given organism, to similarities in, say, skin or bone structure of the same organism, to similarities between ancestors and descendants. Whenever we multicellular life-forms think we're onto a good thing, we want more of it, so we keep copying that good thing until some new condition intervenes to stop us doing so. Where the copying attempts to merely reproduce exactly what has gone before, this can rapidly get ill-adapted and out of control (as in cancer, or rabbit reproduction).[3]

Homology works for living organisms as an energy-saving shortcut. There is no intrinsic reason why change, development, and behaviour in organisms needs to strictly follow past patterns: as we saw in 1.b, incremental change is possible. However, the precise copying of structures that have worked in the past is the most efficient default approach. As long as conditions do not change too much, the probabilities are that the same approach will work again. We do not need to keep 'reinventing the wheel', we just keep re-making the same design of wheel as long as it works. Henry Ford re-discovered the same general point when he pioneered the manufacturing of identical cars on a production line and thus slashed his costs: that is, precise copying is efficient, as long as precisely the same product is required. When living organisms reproduce cells to the same pattern as before, they reduce the 'costs' of matter and energy by avoiding the blind alleys and wastage that variation of any kind would create.

The process of homology is complex, and does not consist merely of genes creating traits like a cookie cutter creating multiple cookies. Sometimes the same genes can have multiple effects – a phenomenon known as *pleiotropy*. On the other hand, traits can be the result of the interaction of multiple genes (*polygenesis*). Different genes can also be responsible for the development of the same trait in different circumstances.[4] Arguably, then, there is a great deal of flexibility in the methods used by organisms to create the reinforcing feedback loops of homology. This complexity builds the likelihood of variation into the process (whether positive or negative in its

2   Roth (1988), quoting Van Valen (1982).
3   See Ellis (2018) 4.a for discussion of three different kinds of creation: ex nihilo, precise copying, and mimesis.
4   Roth (1988) pp. 6–7.

effects), showing another way that reinforcing feedback loops can never be entirely independent of an ever-present chance of change. Nevertheless, the obvious value of homology to organisms in stable environments creates a constant pressure to maintain the similarity of the copying without significant variation.

One of the pressures that helps to maintain the precision of homology in multicellular organisms is that of *specialization*. The more complex a multicellular organism becomes, the more different types of cells it will develop performing different functions. These functions create an environment of evolutionary selection even within a multicellular organism, whereby a cell's failure to conform to its expected function is likely to lead to its elimination. For instance, a white blood cell (*leukocyte*) in the human body that is no longer operating effectively can be killed by specialized types of leukocyte called 'natural killer cells'.[5] There are thus evolutionary niches within multicellular organisms, just as there are in the wider environment, but with a pressure to conform to a specialized role that increases with the complexity of the organism. We will also find the same pattern of specialization and conformity in increasingly refined forms between whole multicellular organisms themselves in both animal and human social groupings. Just as I distinguished in the previous chapter between systemic and absolutized conflict, however, we must distinguish between the systemic specialization that enables multicellular organisms to function, and the absolutized specialization that impedes adaptation.

There are two types of homology identified in biology: *phylogenetic* (resemblances in the structure of related organisms) and *iterative* (reproduction of similar characteristics within one organism). These provide two different routes for reinforcing feedback to potentially become rigidified and run amok, to the detriment of the organism. Phylogenetic homology potentially limits evolutionary adaptation, whilst iterative homology creates the potential for efficient reproductions and specializations within an organism. I'd suggest that iterative homology should also be seen as the maintenance of a particular characteristic unchanged over time: so predation, for instance, provides energy that allows all our limbs and organs to continue in their current structure. A close and constant relationship can thus be traced between homology, or 'copying',

---

5   Vivier et al. (2011).

and reinforcing feedback loops of activity or growth which allow and maintain that copying.

Phylogenetic homology is the basis of reproduction and of genetics in multicellular organisms, so one form of reinforcing feedback loop (reproduction that copies genes) will again need to be offset by the balancing feedback loops of genetic variation and natural selection: I have already discussed this in the previous chapter (1.d).

Iterative homology, on the other hand, has massive advantages for individual organisms. If I have a minor injury to the skin of my finger, I don't want it to be reconstructed with a different type of skin to what went before: rather I want a precise copy so that it is matched to the needs of the rest of my bodily system. Similarly, I have two kidneys, which both have the same function but which back each other up. If one of the kidneys was different in structure from the other, it would not work effectively with the rest of my body. That's an example of the advantages of *bilaterianism*, a specific type of iterative homology which reproduces the same features on both sides of a lateral line along the length of an organism: I will return to that important development in 1.g. Like many of what are initially examples of reinforcing feedback, it further sets the scene for balancing feedback once variation is allowed within the new pattern of copying.

1.f. Nervous Systems, Senses, and Action

> *Summary*
>
> The development of animal nervous systems, used in coordinated sensing and swift, targeted movement, provides a new form of balancing feedback at the multicellular level. Neural processes are also relatively autonomous from those of the rest of the body, separating judgement from the energy used in action, though not to the extent of 'freewill'. However, this new adaptivity also allows new reinforcing feedback loops of predation and environmental modification.

Once cells have clustered together into multicellular organisms, and begun to specialize their roles within those organisms, a problem of coordination begins to arise. The cells can no longer act independently, and the electrical responsiveness of individual cells discussed in 1.b is no longer sufficient to allow the whole multicellular organism to respond to changing conditions. Part of the specialization of cells in multicellular life, then, includes the nervous system, for transmitting information throughout the organism and mobilizing coordinated responses. Other cells also specialize to form sense-organs, to allow information to reach the organism as a whole from outside. A third type of cell begins to specialize in creating the conditions for more effective, targeted action, such as the development of muscles. These three kinds of features are likely to have evolved interdependently, even if at times one may slightly outrun the others.[1] The combination of all three of them seems to be characteristic of animals, even though plants can also be arguably described as having neural systems.[2]

Nervous systems seem to have first developed with the rise of multicellular animals about 600–540 million years ago.[3] Their simplest version is found in ctenophores or comb-jellies. These may or may not have had a common ancestor with all the other metazoans, which went on to develop much more complex nervous systems.[4] Nervous systems are distinct from electrical responsiveness alone in that they also use chemical signals that can be passed from one cell to another to stimulate electrical excitation. The cells of a nervous

---

1   Jekely, Keijzer, & Godfrey-Smith (2015).
2   Brenner et al. (2006).
3   Budd (2015).
4   Jekely, Paps, & Nielsen (2015).

system thus pass on their signals somewhat like a relay race, with chemical signals as the baton, but electrical excitation within the cell as the equivalent of the running between exchanges. This is also aided by nerve cells becoming extended in tree-like shapes, so as to be able to target their signalling to reach other nerve cells.[5]

However, nervous systems within an animal have to evolve in interdependence with the capacity for rapid, targeted external interactions. Animals typically have senses to help them detect external opportunities and threats, nervous systems to mobilize them for action, and muscles to enable rapid movement. Just as the electrical excitation of individual cells is made part of a more complex system by the addition of chemical messages, the senses have developed simultaneously to create such excitation in response to changes in the external environment: whether of light, sound, direct contact, chemical changes in water or air, or even electro-magnetic fields. There seems to be some evidence that all the senses had a common origin in relation to the nervous system, but then subsequently developed different specific types of sensitivity.[6]

The development of muscle, in addition, enables far more targeted and intensive action for multicellular animals than was possible before. Ctenophores (comb jellies) manage to swim using the coordinated actions of cilia (hairs), without muscle, but cnidarians (jellyfish proper) use muscle to enable more vigorous swimming, as well as reaching for prey. These muscles developed through the coordination of cytoskeletons of individual cells into sheets known as *epithelium*.[7]

Once again, then, a move towards reinforcing feedback loops (the development of homology in multicellular organisms) is followed by one of balancing feedback, to allow these new creatures with their greater rigidity of form to counterbalance this with rapid adjustments of behaviour. The nervous system, senses, and capacity for action together create the possibility of modifying our activity in response to new information. Whilst this was also possible for single-celled organisms, the scale and complexity of the response has now taken a step up. This is not an entirely discontinuous step, but one that may have proceeded relatively quickly (for an evolutionary process!) once a tipping point was reached in which

5   Godfrey-Smith (2020) pp. 55–6.
6   Treisman (2004).
7   Godfrey-Smith (2020) p. 57; Keijzer & Arnellos (2017).

the different interdependent elements started to emerge, each incrementally supporting the conditions for the others.

The key feature in the development of the nervous system to be able to respond to new information seems to be its relative autonomy – an autonomy that makes judgement to some extent independent of action. Barandiaran and Moreno's explanation of this relative autonomy is worth quoting at length:

> There is a causal link between the neural domain and certain external processes that belong to other dynamical levels (e.g., metabolic processes in the muscles). This causal connection is largely independent of energetic or material aspects, as neural states produce changes in body states by formal rather than energetic means in a type of lock-and-key causality (what we call formal causality). This causal link is established through the pattern of spikes and not through the energetically determined causality through which these very patterns of spikes propagate. Thus, for example, the motor action caused by neural spikes is not determined by the electrochemical energy that constitutes action potentials but by their form or pattern, which muscle cells 'interpret' (i.e., the process by which the neurotransmitters that neurons generate act by selecting metabolic energy to produce movement). This process is similar to the electric patterns that travel along wires and connect two computers; these patterns produce changes in the terminal not by virtue of the electric energy they convey but by virtue of the sequence of changes in amplitude and frequency.[8]

The nervous system conveys *information*, whilst the metabolic system conveys energy. Obviously the conveying of information still requires energy and is not wholly independent of the metabolic system, yet a tipping point of difference has nevertheless been reached that allows multicellular animals to adapt to their environment in qualitatively different ways. This 'autonomy' is incremental, and continues to develop with the complexity of brains and nervous systems whose responses to new stimuli are increasingly placed in the context of previous responses rather than being dominated solely by metabolic states. There is no support here for totally independent freewill: as psychological studies have found, judges still grant bail more readily after lunch, when there is more glucose in their brains.[9] However, neither are there grounds for determinism, as our judgements become increasingly based on this emergent

---

8   Barandiaran & Moreno (2008).
9   Danziger et al. (2011).

and relatively independent system and their complexity makes them (in practice) impossible to predict.[10]

The tipping point in operational complexity created by nervous systems can be illustrated by comparing the lives of multicellular animals without nervous systems – sponges (*porifera*) – with those who do have nervous systems, such as jellyfish (*cnidaria*). The nervous systems of jellyfish are accompanied by some degree of capacity for sensing and action, though none of the more advanced capacities that will develop in bilaterian animals (discussed in the next chapter). A sponge, on the other hand, is technically an animal but seems very like a plant. After a floating larval stage, it is anchored to one place where it is passively dependent on harvesting passing nutrients by pumping water through its body. Its action potential is limited to stopping and starting this pumping process.[11] Jellyfish in their adult medusa stage, in contrast, have rudimentary nervous systems and light-sensitive organs. They also have muscles that enable them to swim very efficiently, capture prey, and evade capture themselves. A jellyfish is clearly capable of a degree of fast adaptation that a sponge could not manage: for instance, seeking out richer food sources or evading predators.

Although this new responsiveness creates the potential for new balancing feedback loops, of course it also creates the conditions for further reinforcing feedback loops, such as those of predation and competition. Animals with senses and nervous systems can use them to adapt and survive, but they can also use them to exploit their environment ever more effectively. This can be directly observed in the success of jellyfish in taking advantage of anthropogenic disruptions to the ocean ecosystems, such as depletion of fish stocks, eutrophication from agricultural runoff, climate change, and even the erection of structures in the sea.[12] These phenomena have resulted in jellyfish 'blooms', in which jellyfish take advantage of nutrient-rich but oxygen-poor conditions, of a kind that have a negative effect on the fish that might otherwise compete with them. In this case, it seems that the jellyfish's mobility and relative flexibility has enabled it to move into the new niches created by drastic environmental changes, whilst the competition from bilaterian fish is

---

10 VI.3 will discuss the *theoretical* nature of the predictability involved in determinism.
11 Godfrey-Smith (2020) pp. 42–7.
12 Richardson et al. (2009); Munsill (2014).

limited, partly because the fish have more complex needs requiring higher oxygen levels in the water. Although single-celled organisms are of course capable of similar 'blooms', the relative complexity of jellyfish has added a new potential for active and rapid exploitation of a situation.

For the fuller exploitation of these new capacities, however, we need to look to the development of bilaterian animals with bilateralized brains.

1.g. Bilaterianism

> *Summary*
>
> Bilateral symmetry has developed in most animals more complex than jellyfish, and is useful for goal-directed movement. It has also allowed for redundancy that contributes to longer-term sustainability. Bilateral symmetry in the brain, along with cross-wiring to the rest of the body, seems to have first developed in early fish similar to lancelets that lay horizontally in the sea. This is hugely significant for later brain lateralization.

The next decisive step in evolutionary development is that of bilateral organization, which first occurred about 560 million years ago. A bilaterally organized animal is symmetrical in appearance between left and right halves, along an axis that runs the length of the body. This also means that it has a distinct head and tail, as well as a belly and back (ventral-dorsal axis). This bilaterianism is created by homology and is thus immediately an aspect of reinforcing feedback, but it also creates the conditions for many more complex interactions of reinforcing and balancing feedback in the majority of animals, setting the stage for brain lateralization.

Bilaterianism may be dependent on a genetic specification of the left-right axis that affects all multicellular animals,[1] but that was then interpreted and developed in a particular way that only affected bilaterians. Amongst early organisms during the Permian period, the bilaterian group included flatworms and other molluscs, but not comb-jellies or jellyfish. The initial impetus for bilaterianism seems to have been traction,[2] as an animal organized in this way can move forward more efficiently. As Peter Godfrey-Smith puts this:

> *The bilaterian body is set up to go somewhere. There are no non-bilaterian animals at all on dry land – no crawling or walking jellyfish, no anemones with their fingers out in the air.... Bilaterian bodies seem to have begun on a marine version of the land: the seafloor. They are bodies made for crawling over surfaces, with direction and traction.*[3]

Not all bilaterians use traction of this kind against a solid surface, as they include fishes: but the great achievement of fishes is,

---

1   Rogers, Vallortigara, & Andrew (2013) p. 89.
2   Finnerty (2005) also suggests that bilaterianism evolved to improve internal circulation.
3   Godfrey-Smith (2020) p. 72.

instead, rapid and targeted movement through water. Symmetry in general, then, facilitates movement, which is motivated by the goal-orientation of animals. As a further way of fulfilling the goals of animals, this is part of a reinforcing feedback loop.

Some of the earliest bilaterians were fairly simple multicellular creatures like flatworms, which do not yet have a 'through gut', but nevertheless have a head which points forward in the direction of travel, and a mouth. Fairly soon in evolutionary terms, however, most bilaterians developed a digestive tract in which food enters through a mouth and waste is excreted at the other end through an anus. Where the mouth is, the senses also need to be available to find sources of food, leading to the development of a head that combines sense-organs with a mouth.

Other bilaterians, such as the arthropods (including crustaceans and insects), then later vertebrates, developed limbs and other protuberances, such as feelers. Such additional parts, though, tend to follow the same symmetrical model as all other parts of bilaterians: so that lobsters have symmetrical legs, feelers, and pincers, flies have symmetrical legs and wings, fish have symmetrical fins, and humans have symmetrical arms and legs.

One of the key advances in the development of bilaterians is that their genes began to maintain information of the location of a given cell along a head-to-tail axis. These genes, known as *hox* genes, only emerged to operate effectively together in bilaterians, although there are initial traces of them in ctenophores (comb jellies) and cnidarians (jellyfish).[4] Once hox genes had developed for location along the anterior-posterior axis, it was then a relatively small step in evolutionary terms for genes to begin to also specify a lateral and/or ventral-dorsal location. The development of bilaterianism thus created the basis for a variety of types of bilateral asymmetry, as will be discussed further in the next chapter.

Bilateral symmetry is a type of iterative homology (see 1.e). The cells of a multicellular organism are onto enough of a good thing in their development of a given structural form to reproduce it twice, once on each side of the body. In some cases this helps to develop the function of the animal, and thus directly enables it to adapt in a way that helps it to fulfil its needs in its environment. For instance, as already mentioned, bilaterianism aids locomotion, whether an

---

[4] Peterson & Davidson (2000).

organism is moving through peristalsis like a worm, or using limbs like an arthropod or like most land animals. Another example of this is the presence of two symmetrically placed eyes, which can aid the detection of both prey and predators to a greater extent than one eye would do (more on this below).

In other cases, however, bilaterianism produces features that are redundant in the short term. A good example of this is the presence of two kidneys in humans and many other animals. The evidence from healthy living donors seems to be that any immediate negative health impacts of donating one kidney are substantially reduced through time,[5] indicating that the second kidney is to a very large extent actually redundant in the short term. However, as I have already argued in 1.d, redundancy in the short term provides adaptivity in the long term. The kidney performs several vital functions, such as eliminating waste and regulating blood pressure, for which a back-up system may be important within an individual's life-span. This back-up capacity also enables humans to live longer and perform back-up functions in their community that are normally performed by younger people (think of a grandmother, able to continue living though with one defunct kidney, caring for children when their mother is ill).

It is only redundancy of resources that enables adaptivity to a wider range of conditions, so in some ways bilaterianism appears to be a very fortunate development in the complexity of life. In many respects it just provides another new basis for successful competition with other organisms to fulfil needs, another step in the evolutionary arms race. However, in the process of this competitive development, new redundancies are created as a side-effect, and these provide resources for adaptation.

Foremost amongst these potentially fruitful new redundances are those of the brain. The brain seems to have developed in early multicellular animals from ganglia (intersections in the nervous system), particularly those linking the developing senses with motions of the body in response to those senses. Flatworms, for instance, have a simple ganglion situated above the oesophagus that may be regarded as illustrating the evolutionary starting point

---

5   Steiger (2011).

of the more complex brains and central nervous systems of other bilaterian animals.[6]

Bilaterianism also created the two halves of the brain, probably in dependence on paired eyes, each with a separate cluster of nerve connections. The development of paired eyes in early bilaterians can be illustrated in the contrast between two closely related species: the lancelet, a surviving pre-vertebrate ancestor to the fish, has a single eye that is connected to the nervous system through a single *apical organ*.[7] This is the nearest we have to a cyclops! With only a single focus point for sense information, there is no sign of a lateralized brain structure, apart from differing sets of paired neurons in the primary motor centre. Some of these deal with immediate sensory information and rapid responses to it, whilst others deal with longer-term information that has already been processed and is associated with slower swimming.[8] By contrast, a fossil animal called the *haikouella,* very similar to a lancelet, does have paired eyes, leading to the conclusion that it is representative of a sister group to the vertebrates.[9] It seems likely that different laterally placed photoreceptors in the single eye of a lancelet-like ancestor became differentiated into two eyes so as to improve the effectiveness of vision, but in the process the differentiated nerve clusters also became associated with different neurons in the parts of the nervous system controlling movement. Almost immediately in the early development of distinct left and right brain regions, then, we also have an emerging specialization of function. This will become centrally important in the feedback loops of judgement engaged in by all vertebrate life.

From an early stage, too, vertebrates developed the 'cross-wiring' in the relationship between the brain and the rest of the body that humans and all other vertebrate animals are subject to. In this 'cross-wiring', the left side of the brain is strongly connected to the right side of the body, whilst the right side of the brain remains connected to the whole body (though less dominant on the right side). As to why this cross-wiring occurred, and thus also why the distinct linkage between the two different eyes and two different kinds of motor function developed, Rogers, Vallortigara, and Andrew have a

6   Roth (2013) p. 85.
7   Lacalli (1994).
8   Rogers, Vallortigara, & Andrew (2013) p. 67.
9   Mallatt & Chen (2003); Rogers, Vallortigara, & Andrew (2013) pp. 64–6.

probable explanation. Lancelet larvae have an asymmetrical mouth position, with the mouth positioned on the left side of the head, because they spend a lot of time lying sideways in the sea on their left side and gradually sinking, looking for prey as they do so. This 'twist' may also have developed from earlier ancestors that lay on one side on the sea floor.[10] A relationship thus developed between the motor centres on the left side of the head and the mouth on the left.

However, when animals with this ancestry developed two separate eyes, it was the right eye that needed to be connected to this motor centre:

> *The right eye would receive input from prey movements, both when prey was ahead, and when it was to the right. In either case a turn to the right would tend to bring the mouth to bear on the prey; this would be initiated by right eye information, which would have to reach the mouth control mechanisms on the left side of the brain to allow the mouth to be opened in order to seize the target.... Prey detected on the left would need little or no body turning and no particular specialization.*[11]

Thus, it was the importance of the connection between seeing prey and seizing it with the mouth that seems to have created the specific historical link between goal-directed parts of the brain and the right eye (together with the right side of the body more generally) in later vertebrates including humans. By comparison, cephalopods such as octopuses and cuttlefish (which are bilaterians but not vertebrates) do have a specialization between the eyes, but do not have this cross-wiring, left eye being linked to left brain and right eye to right brain.[12]

This evolutionary accident forms a key link in the chain of explanation as to how humans developed a particular tendency towards the short-termist exploitation of the world as essentially 'right', whilst all other considerations were 'sinister' (from the Latin word for left). This, however, is only the beginning of the tale of bilateral asymmetry: one that I shall continue in the next chapter.

---

10 Rogers, Vallortigara, & Andrew (2013), p. 64.
11 Ibid. p. 68.
12 Godfrey-Smith (2020) p. 132.

1.h. Bilateral Asymmetry

> *Summary*
>
> Asymmetry between the two sides of the body of animals is most important in the lateralization of the brain, but many features of the body are also interdependent with brain lateralization. Brain lateralization is specialized and habitual, not essential, and has probably proved advantageous because we can't focus on unique features and instrumental categorizations at the same time, although we need both functions. At group level it also enables rapid coordination.

It would not have taken long after the development of bilaterians for their symmetry to be broken in some respects when this was of clear evolutionary advantage. I have already mentioned the probable ancestor of the lancelet that lay on the sea bed and thus developed a mouth twisted to the left to help it seize prey above. Since cephalopods (octopuses etc.) as well as all vertebrates also have some specialization of brain hemispheres connected to different eyes, even asymmetry of brain structure may have developed in more than one context or go back further than the lancelet. However, as far as vertebrates are concerned, *haikouella*, which lived about 518 million years ago, seems to be a clear candidate for the ancestor of animals with bilateral differences in brain structure. Haikouella was a kind of proto-fish, that helped to define the structure of fishes: a backbone, a streamlined symmetrical body capable of fast and efficient swimming, and two eyes linked to a differentiated brain and nervous system that is able to both find prey and evade predation.

The evolutionary story of the development of further complexity in vertebrates from this point is much more widely known, and it would be little to my purpose to spend too much time on it here. Fish developed jaws and bony skeletons. Some of them crawled onto land and became amphibians, amphibians developed into reptiles, and reptiles into birds and mammals. To continue a very rapid and simplified story, some mammals climbed into the trees and became monkeys, then others came down again and became primates about 85 million years ago. The first proto-human, Australopithecus, developed 3 to 2 million years ago, and had developed into the earliest *Homo sapiens* from about 300,000 years ago.

Homo sapiens, like all other vertebrates, has a basic genetic symmetry, at least on the outside. Irregularities in that symmetry can be troublesome, for instance in problems with mobility created by having one leg longer than the other, or a loss of attractiveness to mates created by asymmetry in the face. Symmetry is in many respects a shortcut indicator of evolutionary fitness.[1] Internally, some asymmetries in the position of the heart and parts of the digestive system do not seem too problematic. However, the asymmetry of the brain is both a major source of short-term adaptation and of a lack of adaptivity in the longer term. This brain asymmetry is also connected in complex ways to the relationship of the nervous system to the rest of the body's motor responses: the dominance of one eye or hand over the other is the most obvious manifestation of this, but humans also have foot[2] and ear[3] preferences, whilst dogs have been observed to have nostril preferences.[4] Greater use of muscles on one side of the body can also change its structure and appearance. Brain asymmetry thus needs to be situated in a wider context of body asymmetry.

The relationship between different types of asymmetry in individuals, however, is complex. One cannot simply identify left-handedness or left-footedness with greater right-brain dominance overall, for instance, although left-handedness appears to be associated with more reliance on certain areas of the right hemisphere.[5] Nor does every individual have the same left-right brain lateralization overall, as there are occasional reversals[6] (to talk of left or right hemispheres, then, is a briefer form for '*normally* left or right'). Nevertheless, left-right brain lateralization is enormously significant, in ways that researchers are still coming to terms with, because it determines the two different *ways* that humans engage with their experience. The figure who has contributed enormously in recent years to the full recognition of brain lateralization and its huge implications is Iain McGilchrist,[7] to whom I refer any reader wanting a detailed account of all the evidence relating to it, as well

---

1 See VII.1 for discussion of the relationship between symmetry and beauty.
2 Schneiders et al. (2010).
3 Bryden (1963).
4 Siniscalchi et al. (2011).
5 Sha et al. (2021).
6 Fischer et al. (1991).
7 McGilchrist (2009; 2021).

as in-depth discussion of its implications across disciplines. In this book I can only offer brief summaries of this material, of a kind that are most relevant to my account of the biological origins of the Middle Way.

It must be recognized that brain lateralization is *specialization* rather than unique functioning. There are many ways that the two hemispheres duplicate functions, in the sense that the other hemisphere can take over those functions following damage to one hemisphere. For instance, those with left hemisphere strokes can often gradually recover at least some left-hemisphere dependent speech functions. However, the functioning of each hemisphere is not virtually interchangeable in the fashion of the two kidneys. Instead, each hemisphere has adopted certain specialized functions. The more it has performed those functions to the exclusion of the other hemisphere, the more it has developed greater efficiency in the performance of those functions, and the more other parts of the body have adapted themselves to this situation. We thus end up with a *habitual* lateralization of functions, the form of which follows the basic patterns set up in our lancelet-like ancestors: namely the left hemisphere for seizing prey and the right for looking out. These two basic functions have developed an increasing complexity through the course of vertebrate evolution, but the patterns are nevertheless extremely clear. The distinction between these two patterns is also that between the two types of feedback loops: the left hemisphere maintains existing patterns whilst the right introduces modifications.

The nature of those distinct functions as they have developed in humans is ably summarized by McGilchrist, drawing in turn on a wide range of neurological evidence. This includes not only evidence dependent on the pathology of people with one injured hemisphere, but also from brain imaging and temporary one-sided disablement of subjects with normal two-sided functioning, plus observation of conditions such as schizophrenia that often reproduce the symptoms of disablement of one hemisphere. Although such studies show us most clearly when the functions of each hemisphere are in isolation, this neurological understanding merely reinforces the multidisciplinary understanding that can also be reached from a variety of other standpoints, as explored throughout this series.

Twenty contrasting features of the brain hemispheres
(adapted from McGilchrist 2021 pp. 28–30)

Left-hemisphere specializations	Right-hemisphere specializations
Manipulation of the world	Understanding and relating to the world
Localized features, detailed and foregrounded	Global features, the whole picture including background
Familiar phenomena	Novel phenomena and anomalies
Certainty, dualistic distinctions	Possibility, ambiguity, provisionality
Less self-critical	More self-critical
Isolated and fragmentary parts, with wholes seen only as mechanically composed of pieces	Gestalt understanding of wholes with complex, integral relationships
Fixity and stasis	Change and flow
Explicit, decontextualized, taking things literally	Contextualized and implicit – thus handles metaphor, myth, irony, tone, and humour
Preference for the inanimate (e.g. tools), with animate things seen instrumentally	Preference for the animate
No understanding of narrative	Understanding of narrative
Categorization using presence or absence of features	Categorization in relation to unique exemplars (e.g. prototypes)
More general and abstract categories	More fine-grained categories and recognition of individuals
Tends to focus on decontextualized body parts as objects	Awareness of internal body image, and of body language and emotion in others
Fine analytic sequencing, linguistic vocabulary, and syntax	Overall meaning of language. Rhythm, tone, and inflection in language
Simple rhythms and abstract musical analysis	Most musical awareness for most people
No understanding that others have a point of view	Understanding of others' independent point of view
No empathy	Empathy
No emotional receptivity or expressiveness	Emotional receptivity and expressiveness
Representation (abstracted signs)	Pre-conceptual recognition, direct experience
Unreasonably optimistic and unaware of own limitations	Realistic and inclined to pessimism

The table offers a version of McGilchrist's summary of the specialized features of each hemisphere. To see these in the context of the biological developments I have been charting, it is necessary to reflect particularly on the ways that identifying and seizing prey requires a consistently decontextualized, instrumental mode of operation. The prey of earlier animals has developed in the context of human life into goals of all kinds – goals that have to be separated out from their context and instrumentally sought to the exclusion of other considerations. We cannot allow distractions from our goals, so the overall mode when we are focused on them is certainty, not investigation or questioning. We may devise a sequence for reaching our goals, but our progression through that sequence tends to be marked by impatience, not by an acceptance of the experience of time passing. Our prey is also wanted for its parts, its muscle and entrails that we want to devour, not for appreciation as a whole. The focus on catching prey thus accounts for a wide array of features of left-hemisphere thinking that we might now associate with the goal-driven norms of, for instance, business, government bureaucracy, or private exploitation of others.

In contrast, it is the right hemisphere that deals, instead, with the wider balancing context of these goal-driven activities. Not only is the right hemisphere on the lookout for possible threats that may come from any quarter, but it also provides the wider appreciation of meaning that makes our goal-driven predation worth pursuing in the first place. It is the right hemisphere that monitors our awareness of the body from within, a faculty that has developed into our awareness of emotion and of the wider significance of our concepts and other symbols. It is the right hemisphere that also handles our relationships with others – primarily of our own species, but also of other animals that we view as companions rather than as prey. Our awareness of their independent viewpoint, and our ability to recognize and respond to their emotional states is a right-hemisphere faculty that can be overridden if we switch to the instrumentality of prey-seeking.

That there is a trail through evolutionary time leading from the lancelet-like ancestor to the modern human condition has been confirmed by a range of studies on vertebrate animals which confirm the continuing effects of brain lateralization. This is evident not only neurologically, from experiments on animals under controlled conditions,[8] but also from their everyday behaviour. It is thus possible for non-expert everyday observers to note the effects of lateral

---

8   Rogers, Vallortigara, & Andrew (2013) pp. 3–27.

asymmetry in animals: for instance, close observation of birds in your garden is likely to confirm expert observations such as these.

> Fish keep track of their companions using by preference their left eye and of a predator using their right eye; lizards, toads, baboons and wild horses are more likely to attack a conspecific competitor seen by the left eye; chickens and Australian magpies look at aerial predators using their left eye; black-winged stilts...use their right monocular visual field before predatory pecking, whereas shaking behaviour, a component of courtship displays, and copulatory attempts by males are more likely to occur when females are seen with the left monocular visual field.[9]

That asymmetry has evolved and continued in so many animals clearly suggests its overall functional value for them. As its authors point out, the animal behaviours in the quotation above are not obviously evolutionarily advantageous, and on the face of it appear disadvantageous.[10] Predators may appear from the right as much as the left, and prey from the left as much as the right. Moreover, predators could potentially learn to take advantage of the vulnerability involved in their prey having less open attention in relation to the right eye, in the same way that they might learn to stalk them from downwind. As we will see, these vulnerabilities associated with asymmetry can be readily transferred to the human context. Asymmetry only becomes helpful in fulfilling an organism's needs when its functional advantages outweigh its disadvantages, and Rogers, Vallortigara, and Andrew suggest two ways in which it might do so.

The first of these advantages lies in the idea that the functions of the two brain hemispheres may actually be incompatible: that is, a single undivided system either could not (or could not efficiently) perform both of them at the same time. Yet, survival may often depend on them both being performed at the same time. The incompatible features of the two different functions are those of paying attention to both unique and comparable features (variance and invariance) in any new phenomenon. As Rogers, Vallortigara, and Andrew explain:

> When assessing a novel stimulus, an ability needed quite commonly by even the most primitive vertebrates, two different types of analyses must be carried out. First, the organism must rely on previously comparable experiences to estimate

---

9   Ibid p. 39.
10   Vallortigara & Rogers (2005).

> the degree of novelty of the stimulus, and to do so it must recall shared memories and then elaborate on them for future use. This process requires attention to the *unique* features of an event or stimulus (i.e. the variance, outside the range expected from previous experiences). Second, certain appropriate cues, based on past experience and/or phylogenetically based instructions, must be used to try to assign the stimulus to a category, and so to decide what sort of response (if any) should be given. Categorization must be made on the basis of selected stimulus properties, despite variation in many other properties.[11]

This seems to me to also accord with human experience. We cannot focus on the *unique* characteristics of an object at the same moment as we categorize it abstractly or instrumentally, but rather, at best, we tend to flip between them. We can experience the aesthetic beauty of a flower and we can classify it botanically – but not at exactly the same time. We can also empathize with another person, on the one hand, or regard them as an economically or sexually exploitable object, on the other – but not at exactly the same time. We have one channel of attention, but two available systems connected to it.

The second advantage of asymmetry lies at the population level: that is, that the coordination of asymmetry in groups of animals creates more efficiency in the entire operation of that group, at least in relatively stable conditions.[12] There is a variety of degrees of asymmetry amongst all the individuals in a given population of animals, and the optimum degree of asymmetry follows a bell curve.[13] However, pressure for a continuing degree of asymmetry may be maintained by the need for conformity in a wider group in turn aiding the chances of survival for each individual. This limited advantage depends on the relative predictability of any individual's behaviour under left-hemisphere dominance, as it is the left hemisphere that uses prior models to determine behaviour, and thus regularizes behaviour in a group. The coordination of fish in a shoal, or birds in a large flock, depends simultaneously on them being aware of each other's movements (right hemisphere) and being coordinated to respond in similar ways (left hemisphere). Researchers have successfully used computational algorithms to track collective

---

11 Rogers, Vallortigara, & Andrew (2013) p. 46.
12 Ibid. p. 52.
13 Ibid. p. 53.

behaviour in animals, suggesting that such behaviour follows the left-hemisphere models that can be translated in this way.[14]

This is an advantage in relatively stable conditions because of the ways that it rapidly multiplies reinforcing feedback loops, enabling mass copying of the same activity by many individuals of the same species. A pod of dolphins can thus coordinate their hunting of a shoal of fish, and the shoal can also attempt to resist through collective movement. A plague of locusts can illustrate the devastating environmental impact such coordination can have amongst animals – although such rapidly multiplying, coordinated animal groups can also quickly destroy the very basis of sustenance that made their coordinated activity possible. Perhaps for this reason, there is also evidence from research in fish shoal behaviour that individuals within an animal group can display a variety of degrees of lateralization, and that this may be beneficial for the group as a whole.[15] For instance, less lateralized fish swimming on the edges of a large shoal may be better able to look out for predators than their more conformist peers in the middle of the shoal.

As we will see in the next chapter, this second advantage of lateralization has important implications for its effects on humans, creating a social and cultural pressure for not just brain asymmetry, but the dominance of one asymmetrical hemisphere over the other. Asymmetry alone, between balanced and coordinated hemispheres, allows the reinforcing feedback loops of our self-assertion to be balanced by the wider contextualization offered by the right hemisphere. However, the chronic dominance of one hemisphere over the other creates an unparalleled potential for self-destruction.

---

14   DeLellis et al. (2014).
15   Killen et al. (2017), p. 6.

1.i. Left-hemisphere Repression

> *Summary*
>
> Absolutization is the specifically human form of reinforcing feedback that develops due to left-hemisphere repression of the right, via the corpus callosum. Human brains are more lateralized as a result of being bigger, leading to a greater possibility of reduplicated functions, and thus a greater practical need for suppression or repression of each hemisphere by the other. Language and cultural reinforcement building on group coordination, though, makes left-hemisphere repression of the right dangerously powerful.

After tracing the succession of reinforcing and balancing feedback loops in the story of the development and evolution of life as a whole, we have now finally reached the point where reinforcing feedback loops take a distinctively human form. That form consists in the development of absolutization, the phenomenon that I discussed in depth in the first volume of this series.[1] As I discussed in that book, left-hemisphere over-dominance is one of the dimensions of absolutization, even though one can also approach it in many other ways. It is left-hemisphere over-dominance (or left-hemisphere repression), however, that particularly marks the development of absolutization in the evolution of living organisms. As I also suggested in that book, absolutization can be understood more generally (and somewhat informally) as the belief that we have the whole story.

Most other animals share brain asymmetry with us, as discussed in the previous chapter, and any animal with brain asymmetry can potentially get 'stuck' in a left-hemisphere mode, resulting in fruitless and repetitive behaviour. For instance, dogs can get obsessive-compulsive disorder in which they fixedly guard something or chase their own tails.[2] This could be seen as an implicit or partial form of absolutization. However, only humans, as far as we can be aware, hold potentially explicit beliefs that they have the whole story. When we do this, we are immersed in the certainty and fixity of the left hemisphere, to the exclusion of the functions of the right.

---

1   I:1–5.
2   E.g. https://www.doglistener.co.uk/obsessive-compulsive-disorders-dogs (accessed 2022). The lateralization of such complaints in animals seems to be little researched. See McGilchrist (2021) ch. 9 for very detailed discussion of the lateralization of some forms of mental illness in humans.

The major development of the pre-frontal cortex (in both hemispheres) in primates, followed by the human development of language that we take to be representational, has massively increased our capacity for absolutization compared to that of other animals. Of the 23 dimensions of absolutization that I discussed in my earlier book,[3] I would estimate that only three are likely to occur in other animals – namely reinforcing feedback loops, repression and conflict, and group binding. Dogs may get into obsessional states, but they do not create elaborate accounts of 'reality' to rationalize them.

The biological story as to how absolutization developed through left-hemisphere repression has been convincingly told by Iain McGilchrist. This story is one of the side-effects of primate (and then specifically human) brain development, and of the inhibitory processes that we need to be able to operate with such big, complex brains. Firstly, brain lateralization has reached an unprecedented apogee in humans. Secondly, our increasingly specialized brains have created reduplication and overlapping of functions which then requires an increased capacity for inhibition. When our capacity for apparently representational language is added in early human cultural development, the stage has thus already been biologically prepared for rigid beliefs and their socio-political exploitation as a side-effect of human intelligence.

The first stage of that story is that although all vertebrate brains are lateralized, some are more lateralized than others, and human brains are supremely lateralized. The reason for this seems to lie in the mathematical relationship between the number of points (neurons in this case) and the number of relationships needed between those points, such that as numbers of neurons grow arithmetically, numbers of connections need to grow exponentially to keep those neurons connected into a single system.[4] As primates and then humans have developed the neocortical areas of the brain, then, the connections have struggled to keep up to maintain the effectiveness of the system, despite the development of myelin sheathing for more effective connection. To make the system manageable in its complexity, increasing specialization is the only available option. Thus the brains of primates have been observed to increase in specialization compared to those of rats, just as the pre-frontal cortex

---

3   I: Appendix.
4   McGilchrist (2021) p. 59.

has increasingly developed.⁵ This pattern, a general characteristic of complex systems, has of course also been reproduced at a social level in modern human economies, where the difficulty in transmitting increasingly complex information to everyone has resulted in increasing specialization and the delegation of whole areas of our understanding to experts. As the human brain has grown in complexity, then, both hemispheric lateralization in general, and other forms of specialization within the hemispheres, has grown.

Secondly, then, in response to this increasing specialization in the human brain, comes the re-duplication of functions. Given the general symmetry and re-duplication of some functions between hemispheres, many stimuli are likely to excite *both* hemispheres initially.⁶ To have both hemispheres contributing to our response may increase its effectiveness in many cases: as discussed in the previous chapter, they can offer two different complementary kinds of attention, focusing either on variance or invariance. However, to build up an adequate set of beliefs that then enable action, those different kinds of hemispheric activity cannot be allowed to interfere with each other and cause confusion. Suppression, or repression, of one hemisphere by the other is the result.

Hence the role of the corpus callosum connecting the hemispheres. The corpus callosum is not the only connection between the hemispheres, but it is the main connection at the higher, newer levels of the brain, as opposed to the brainstem. The role of the corpus callosum, however, seems to be at least as much inhibitory as conductive.

> When it comes to sensory and motor information, connectivity between homotopic regions (the 'mirror image' regions on either side) serves to co-ordinate the hemispheres in action; but where the most cognitively complex activity is going on, in the heteromodal association areas, communication is less important – unless it is actively to **suppress** one side or the other.⁷

Either hemisphere can inhibit the other in order to enable its own mode to take over when we need to operate in that mode. However, the two hemispheres inhibit in slightly different ways – a distinction that seems to be behind the difference between suppression and repression. Whilst suppression from the right pre-frontal cortex

---

5   Uylings & Van Eden (1991).
6   McGilchrist (2021) p. 62.
7   Ibid. p. 60.

gives us inhibition of left-hemisphere processes *with awareness of a wider context*, repression from the left pre-frontal cortex, in accordance with the whole mode of left-hemisphere operation, limits our awareness to that of the left hemisphere's represented beliefs.[8] Suppression of the left hemisphere by the right is thus helpful for provisionality, enabling us to avoid being dominated by one set of assumptions whilst we consider alternatives. Repression of the right hemisphere by the left, however, is the source of absolutization. We make judgements in ignorance of any alternatives, because awareness of any such alternatives has been blocked.

Such repression could be relatively harmless if it lasted a short time, used for the purposes of action but preceded and succeeded by right-hemisphere awareness. In that case the left hemisphere's tendency to isolate the current moment of judgement from any wider awareness of time[9] could be used to good effect, because sometimes we need to restrict awareness for purposes of action – for instance to escape from danger, to save someone else from imminent danger, or to procure vital food. However, left-hemisphere repression becomes increasingly problematic the more it is maintained beyond such vital moments of action, becoming the source of proliferation in human mental states[10] through reinforcing feedback loops of decontextualized belief. The socio-political role of absolutization seems to have also been crucial in the development of left-hemisphere repression as humankind's largest problem.

As mentioned in the previous chapter, the coordination of brain asymmetry in groups of animals seems to have been one of the causes of its continuing development in vertebrates. Where such coordination is primarily that of the right hemisphere, this will have the effect of helping organisms in groups respond to each other as well as to their wider environment. In humans, the term *solidarity* seems to capture this: we cooperate because of our awareness of each other, not because of any unconditional sharing of belief. However, where such coordination is dominated by the left hemisphere, with the awareness of the right repressed, it provides a shortcut to group cooperation that shuts out wider awareness – one that becomes increasingly problematic the longer it lasts and the more disconnected it is from wider conditions.

8 I.5.a; II.2.e; McGilchrist (2021) p. 223; Kinsbourne & Bemporad (1984).
9 I.3.c.
10 I.1.a.

The development of language that we take to be representational (when it is understood only in the terms of the left hemisphere) offers propositional markers that both unite the beliefs of people in a group, and maintain those beliefs over time as they are repeated by different people in the group to each other. Left-hemisphere repression thus comes to serve group purposes, stimulated by statements of shared belief, and uniting the group in the service of those shared beliefs. Shared group representations constantly interact with left-hemisphere dominance in individuals, so as to reinforce it. Left-hemisphere repression simultaneously creates unnecessary conflict both at group and individual level, as inconsistent beliefs can no longer be reconciled through investigation. This group conflict has its correlates within individual psyches, where the represented voices of others conflict with each other, and recruit our motives inconsistently at different times. Each side of such conflict insists that it has a representation of 'reality', and develops elaborate representations of that 'reality' – metaphysics is born.

At this point in the history, however, biological conditions in the development of absolutization are starting to be succeeded by cultural ones. I will be looking at these in more detail in section 3 of this book, beginning with the roots of cultural absolutization in the paleolithic and neolithic periods of human history. Before I continue that story, however, there is one more aspect of the biological story to consider – namely that of the biological capacity to overcome absolutization. Left-hemisphere repression is never the whole story. Integration of the hemispheres is also part of that story.

1.j. The Biology of the Middle Way

> *Summary*
>
> Absolutization is due to conflicts between representations in the human left-hemisphere pre-frontal cortex at different times when judgement is applied, but parallel capacities to reframe and contextualize these absolutizations have also developed in the right pre-frontal and parietal cortices. The Five Principles of the practice of the Middle Way can all be given neural explanations that unite these capacities of the right hemisphere with support from the left. This has the effect of re-uniting the energies experienced before the conflict emerged, though we also have more to integrate because of having gone through that process.

The biological conditions have left us with a succession of evertighter reinforcing feedback loops that are then reinforced by human culture to produce absolutization. The negative effects of absolutization are profound, as discussed in volume I of this series, and they culminate in the increasing risk of self-destruction by human civilization. However, the biological story is not all negative, because those same conditions that have produced absolutization in the context of reinforcing feedback loops, have also produced a succession of balancing feedback loops. The final balancing feedback loop (in the ascending complexity so far) is that of integration and the Middle Way, which is the positive theme of this whole series.

Integration is the overcoming of conflict between opposed desires and beliefs (and less directly, the defragmentation of meaning), at both individual and socio-political levels.[1] In 1.c above I distinguished between *systemic conflict*, which is an unavoidable part of the functioning of a wider system, and *absolutized conflict*, which is not practically necessary to the functioning of a system, but instead arises from the continuation of reinforcing feedback loops in a non-adaptive sub-system. Another animal could conceivably be 'stuck' in a way that involves absolutized conflict, after the fashion of an obsessive-compulsive dog, but on the whole absolutized conflict is a human speciality. The resolution of systemic conflict can occur just through a change in conditions – for instance, if the American grey squirrels that have become dominant in the UK reproduce less, then they will be in less conflict with the native red squirrels. Humans can also be relieved of conflicts just because the context for that conflict in competition or predation has changed – for instance tastier vegan foods become available in the shops, making less eating of animals desirable. The integration I am concerned with here, though, is that which resolves absolutized conflicts.

Absolutized conflict is not resolved through the dominance of one value over another understood at the level where the conflict occurs. Rather, it requires *reframing* so as to create a different representation of the conditions in which there is no longer a basis of conflict. Instead there is just a wider set of conditions to be addressed, including those previously addressed by both sides of the conflict from a more limited point of view.[2] That reframing may show one side to be addressing conditions a great deal more than the other, but it is nevertheless not unconditionally accepted.

This integration is bilateral in the sense that it overcomes conflicts that have been created by left-hemisphere repression of the right, and resulting absolutization. However, crucially, it is not integration *of* the right hemisphere with the left, but integration of conflicts between beliefs held in the left hemisphere at different separated times. It is only the right hemisphere that can create the wider awareness that both continues to acknowledge the conflict and then contextualizes it. The left hemisphere, in its own terms, is not capable of integration, because it is blind to conflict that depends on information external to its prior assumptions held at a specific time. Left-hemisphere assumptions held at different times, or by different people, constantly conflict in ways that require the *Middle Way* to be used as a basis of belief. Hence the integration process to overcome absolutization is one created *by* the right hemisphere, with the left hemisphere's role limited to cooperating with the right's wider perspective. At each moment of judgement when the Middle Way can be applied, it is applied because of the presence of right-hemisphere awareness.[3] It is thus the 'middle' way between our opposing beliefs when we formulate them only with the left hemisphere, not the 'middle' way between the hemispheres.

So what is the biological basis of this capacity for integration to overcome absolutization, and how did it develop? In *The Five Principles* I distinguished five elements of Middle Way Practice, all

---

2   II.5.b.
3   See also I.5.a; II.5.a.

of which are required for effective long-term practice to overcome absolutization.[4] All of these, I think, can be understood as capacities of the human brain when the right-hemisphere capacity is taken sufficiently into account (as well as 'principles' for deliberate practical application). The five principles are scepticism – the practice of recognizing uncertainty in our beliefs, provisionality – the practice of considering alternatives to absolutes and their mere negations, incrementality – seeing things as a matter of degree, agnosticism – the maintaining of Middle Way perspectives in the face of dualistic pressure, and integration – the uniting of desires, meanings, and beliefs both in the shorter and longer term, both at individual and at socio-political levels.

The first of these, scepticism, depends on our capacity to process incomplete information. The awareness of uncertainty, that is the awareness that we do not have complete information, has been clearly shown to lie in the right pre-frontal cortex rather than the left.[5] This would be exercised, for instance, if we respond to someone's view, expressed on social media, with which we disagree – but at the same time we need to realize that we know almost nothing about this person, their mental states, social situation, or other views. Only the right hemisphere can deal with the ambiguity of unresolved information, rather than treating it as true or false, and it is the pre-frontal cortex that overwhelmingly supplies our critical awareness.

The second principle, provisionality, involves awareness of a wider context, together with suppression of the sole dominance of absolutized beliefs or their negations. As already mentioned in the previous chapter, this suppression can be distinguished from repression because it involves a shift of attention *due to* wider awareness, rather than shutting out that awareness. For instance, we might modify our response on social media *because of* an awareness of how little we know about the other person, contextualizing our initial feelings of opposition, and thus suppressing our initial impulse to an aggressive response. This faculty of suppression is also located in the right pre-frontal cortex. In addition, the right pre-frontal cortex is closely involved in the process of 'weighing up', as I discussed it

---

4   II.1–5.
5   Goel et al. (2007); McGilchrist (2021) p. 274.

in *The Five Principles*, that is, making provisional judgements about the wider options that have been brought into awareness.[6]

The development of the right pre-frontal cortex, then, has been crucial for the development of the human capacity for Middle Way judgement in both scepticism and provisionality. However, the left and right areas of the pre-frontal cortex have evidently evolved simultaneously: just as the over-dominance of the left pre-frontal cortex has created absolutization, the contextual remedy for it has also been available to us as a result of the same processes. The pre-frontal cortex in humans is considerably larger than that of other primates, but this also seems to be less significant in itself than the extent of white matter in it, which determines the complexity of links.[7] Teffer and Semendeferi, in a review on the evidence of pre-frontal cortex evolution, write

> *Conventional wisdom holds that there is a grade shift in cognitive abilities between humans and extant great apes,... suggesting that many traits are the sole province of human intelligence, and evolved sometime after chimpanzee and hominin lineages diverged from a common ancestor. At this time, it seems that evolution in human ancestors was accompanied by discrete modifications in local circuitry and interconnectivity of selected parts of the brain.*[8]

Human evolution thus seems to have enabled *both* absolutization in the left hemisphere *and* the capacity to contextualize it through developments in the right. We develop that capacity further by using it, which increases the density of the neural links of meaning, building on the already complex interconnectivity of the human brain.

The third principle, incrementality, seems to depend much more on our capacity to maintain awareness over a duration of time. As movement either of attention or of our bodies through space requires temporal continuity, our sense of temporal duration is also required for continuity of tracking in the objects we examine.[9] For example, to recognize that a person has a *degree* of trustworthiness rather than being just totally trustworthy or untrustworthy, we need to track them over time with continuing contextualized awareness of their variations in trustworthiness. This awareness over time

---

6   II.2.g; Stolstorff, Vartanian, & Goel (2012).
7   Teffer & Semendeferi (2012).
8   Ibid. p. 210.
9   II.3.f.

is particularly located in the right parietal cortex, in connection with the right temporal and frontal lobes[10] and can be contrasted with the left hemisphere's tendency to segment experience of time into discontinuous sequences. Paleoneurology suggests a substantial expansion of the parietal regions in humans when compared to their ancestors.[11] The expansion of the left parietal area is associated with the coordination of tool use, but the right much more with the coordination of wider temporal and spatial awareness.

The fourth principle, agnosticism, is probably alone amongst the five principles in having a more left-hemisphere focus, but this is indicative of the ways that the two hemispheres need to work together in all integrative practice. In contrast to scepticism, provisionality, and incrementality, agnosticism has a much more defensive function. It involves acknowledging that our embodied human situation does require us to maintain positions of belief and justify some positions over others, albeit positions that are maximally justified in their complexity, breadth, and provisionality. Once we have established that the Middle Way does involve some positive positions and commitments, agnosticism helps us to maintain a critical consistency in our commitment to the Middle Way, without which it may readily be appropriated back into absolutized positions of the kind that assume that the Middle Way is actually on 'our' side or on 'their' side. This requires a connection between the skill of comparative or relational thinking located in the right pre-frontal cortex and exercised in scepticism, together with the kind of categorical and propositional thinking based in the left pre-frontal cortex. Both of these, again, are obviously distinctively human processes. The relational thinking can make us aware that the Middle Way is something distinct, involving reframing of the conflict, in the first place, but the propositional thinking is also required to offer a clear and consistent alternative position.

The final principle of the Middle Way, integration, depends for its reframing of beliefs on the other four principles. However, the integration of desire as its most immediate form is simply the unification of energies that were previously divided, aiding the conditions for provisional thinking and feeling, as well as representing their effects. That unification of energies in some ways parallels

---

10 McGilchrist (2021) pp. 78–9 integrates a wide range of neurological sources on this.
11 Bruner, Battaglia-Mayer, & Kaminiti (2022).

the integration of energies *prior* to the conflicts created by absolutization, and thus present to a greater extent in both animals and young children. The recognition of this point is found, for instance, in the belief that babies in the womb are enlightened,[12] or in the poet Rainer Maria Rilke's insights into animals:

> *Did consciousness such as we have exist*
> *In the sure animal that moves towards us*
> *upon a different course, the brute would drag us*
> *round in its wake. But its own being for it*
> *is infinite, inapprehensible,*
> *unintrospective, pure, like its outward gaze.*
> *Where we see Future, it sees Everything,*
> *Itself in Everything, forever healed.*[13]

In one sense, this does indeed show integration occurring both phylogenetically and ontogenetically before the adult human state of being subject to absolutization. As we have seen, however, left-hemisphere conflict does not start only with human beings, but is potentially a wider problem for all vertebrates – especially those with relatively greater levels of lateralization and neocortical development, such as other primates, or dogs. Perhaps an insight into this is what prompts Rilke's next stanza, which immediately follows on to qualify the above.

> *And yet, within the wakefully-warm beast*
> *there lies the weight and care of a great sadness.*
> *For that which often overwhelms us clings*
> *to him as well, – a kind of memory*
> *that what we're pressing after now was once*
> *nearer and truer and attached to us*
> *with infinite tenderness. Here all is distance,*
> *there it was breath. Compared with that first home*
> *the second seems ambiguous and draughty.*
> *Oh bliss of tiny creatures that remain*
> *For ever in the womb that brought them forth!*
> *Joy of the gnat, that can still leap within,*
> *even on its wedding day: for womb is all.*[14]

Such integration is not a resolution of *systemic* conflict. The gnat will still be snapped up by a passing swallow. What is more, of

---

12 E.g. discussed by Trungpa (2002) pp. 151-2.
13 Rilke, trans. Spender & Leishman (1939), 8th Duino Elegy.
14 Ibid.

course, it is not a result of a process of overcoming absolutization. It is thus not integration in the sense of the principle of the Middle Way. To conflate this relatively integrated state prior to absolutization to one achieved through the practice of the Middle Way is an example of what Ken Wilber called the *pre-trans fallacy*.[15] The *capacity* for more integrated states goes back phylogenetically in time and also ontogenetically (as we will see further in the next section) to infancy, but the process of developing them, and hence the value of the experience of having them integrated, does not. It is only by reframing that we get the benefits of integration, because prior to that we do not have so much to integrate.

Overall, then, in biological terms the Middle Way can be seen as overwhelmingly the product of distinctively human capacities in the neocortex – ones that have been responsible for causing the problem as well as offering the solution. However, just as absolutization is built on a whole sequence of previous developments that created new reinforcing feedback loops – from the maintenance of life itself to the developments of homology, reproduction, and bilaterianism – so is the capacity for the Middle Way built on the sequence of previous balancing feedback loop developments that have allowed us to respond and adjust to our environment: electrical responsiveness, the development of the senses and nervous system, and the capacities of the right hemisphere. The Middle Way, without all these other kinds of potential balancing processes beneath it, would not be able to protrude, iceberg-like, above the surface. Without these biological developments in previous organic lives, too, it could not appear in individual lives or in cultural development, the themes of the remainder of this book.

---

15 Wilber (1982).

2. Stages of Psychological Development

Robert Kegan's Theory of Stages of Psychological Development, as discussed in section 2

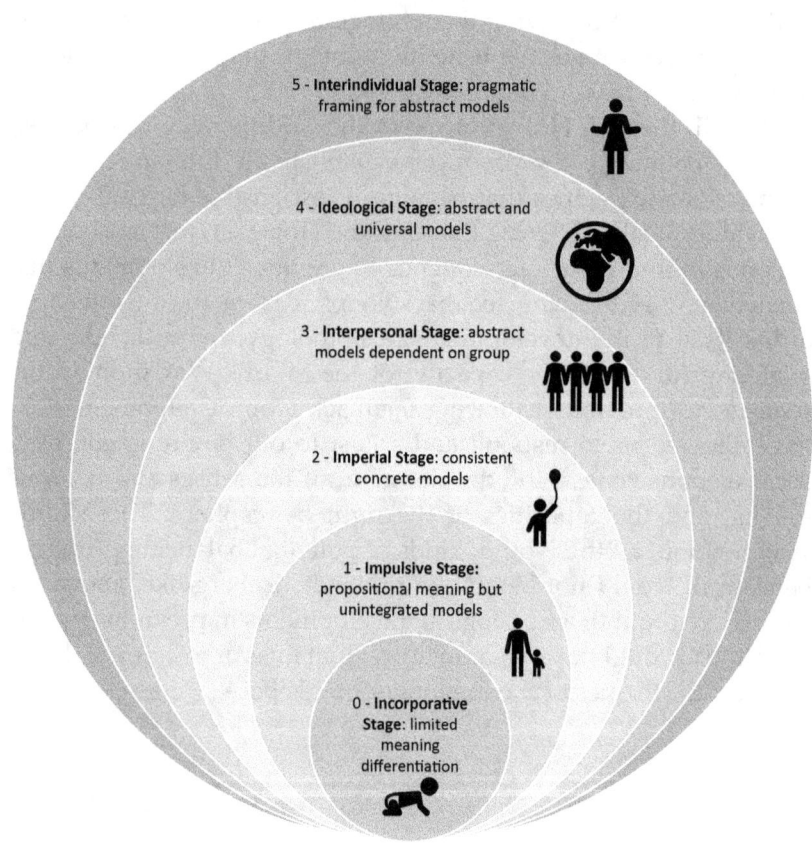

2.a. Issues of Psychological Development

> *Summary*
>
> The approach in this book favours the psychological development theory of Robert Kegan, because this integrates various approaches, and identifies tipping points of judgement as the basis of stages. Although a particular culture has a crucial role in enabling individual development to reach a particular level, it becomes speculative to stage cultures as a whole. Stage theory is a helpful tool used provisionally, but we also need to avoid using it for any purposes of social power: this would absolutize what is only one aspect of our asymmetrical characters.

The Middle Way is a principle of *judgement*, not of evolution nor of general character development. Nevertheless, the story of how we reached the point at which we need to make our judgement at each moment is of some relevance to making that judgement as well as we are able. In section 1 I have focused on the biological and evolutionary story of how humans in general have reached a point where the Middle Way is needed. This perhaps is of some practical help in the way that it helps us to understand the biological conditions we have to address in our judgements. However, the bigger context that will probably loom far larger for any given specific judgement is that of our personal story. How did I, as an individual, get to the point where I need to respond to this particular situation? This story is that of my *development* as an individual – in biological terms, of my ontogeny.

The story of each person's psychological development is only partly a biological story. It is shaped by our shared genes and other shared conditioning as human beings, and also by epigenetic processes through which ancestry may have shaped the expression of genes. It is, however, shaped just as much by our socio-cultural context. Our genetic and epigenetic heritage may not reach its full potential to shape our capacities if we are not also appropriately stimulated by our relationships and cultural environment. Sections 3 and 4 of this book focus on how that cultural environment has developed. In addition, our past judgements increasingly shape the context of our ongoing judgements as we grow older: for instance, a decision to study at university, and the subject we choose, has a huge shaping effect on the rest of our lives. All of these kinds of conditions need to be taken into account, as far as we can manage it, in the contextual assumptions we make whenever we make judgements.

Theories of psychological development attempt to describe in general terms the stages to which our common biological heritage predisposes us as humans, provided we receive the equally necessary socio-cultural stimulus. I would argue that there is no necessary conflict between theorists who have emphasized the biologically universal basis of the stages, such as Piaget, as against those who have emphasized the socio-cultural element, such as Vygotsky, so long as we continue to recognize the continual interaction of both elements. These also interact with the third factor, of individual past choices. Psychological stage theories are extremely useful tools, but must always be treated provisionally as ways of helping us understand the context in which a person makes a judgement. They do not tell us exactly what that judgement will be in advance, but their limitations should not block our understanding of the substantial ways that they can help us.

Psychological stage theories are potentially very helpful as tools for incrementality of training, as discussed in *The Five Principles*.[1] Absolutization in our expectations either of ourselves or others is liable to take the form of a discontinuous assumption either that they will maintain current habits of judgement, or that they will suddenly be able to meet a new ideal. In the former case, we are stuck, and in the latter, we are 'in over our heads' as Robert Kegan puts it.[2] However, advances in judgement begin from where we are now and stretch us slightly from there. If we can familiarize ourselves with the main psychological stages, they can provide a starting point for appropriate expectations. They are a tool to help us focus on *progression* rather than only the goals of progression. In that sense this section may have a practical application that the discussion of biological evolution in the first section will have lacked.

However, psychological stage theories are often controversial, because they raise various problems, and are also easily open to abuse. The most basic problem is that of how and why to identify stages. Is human development not a continuity? Yes, it is a continuity, but *with tipping points*. As I argued in *The Five Principles*, the recognition of tipping points is entirely compatible with a commitment to incrementality, as incrementality tells us nothing about

1 II.3.g.
2 Kegan (1994).

the speed of change, and tipping points are just rapid continuous changes.[3] The basic principle of a successful stage theory, then, is that it should successfully identify tipping points, which are universal because they are an outcome of general human conditions. Psychological tipping points involve changes in priority when we make judgements – in other words changes in value. We do not instantaneously change from one value to another, but we become aware of new options, and these gradually become prioritized in the judgements we make.

Different stage theories have focused on different ways of determining stages and their tipping points, in accordance with established boundaries of discussion. There have been stage theories based on psychosocial relationships (Erikson),[4] stage theories based on cognitive development (Piaget),[5] and stage theories based on moral development (Kohlberg).[6] For Erikson, each stage has a characteristic psycho-social task, in which we learn to relate to others in slightly more mature ways, and thus overcome internal or external conflicts at the previous level. The tipping point, then, comes when we gain confidence in a new habitual way of relating, and this rapidly supplants the old. For Piaget, each new stage involves a schematic system of manipulating symbols and concepts at that level, which is maintained as long as it functions effectively, until *disequilibrium* causes first flips and then new contextualization of that level of framing into a higher one.[7] For Kohlberg, increasingly universal and flexible moral principles are applied at a succession of stages, again as a result of conflicts or disequilibrium in the previous stage.

The structural similarity and interdependence of different stage theories should thus be clear from the common pattern in these three types of influential stage theory. Psycho-social relationships, 'cognitive' development, and moral development should not be treated in isolation from one another, but rather seen as differing perspectives on personal development. These features can also be understood as the progression of long-term integration that I discussed in *The Five Principles*, connecting integration of desire, meaning, and belief,[8] and culturally supported by the virtues cultivated in communities

3   II.3.b.
4   Erikson (1982).
5   Piaget (1955).
6   Kohlberg & Hersh (1977).
7   I.2.c.
8   II.5.d.

of practice.⁹ Whatever stages of progression we identify, the basis of doing so should be clear tipping points created by the integration of conflict at the previous level, with reframing¹⁰ and contextualization at the next level.

These tipping points also mark a shift from reinforcing to balancing feedback loops, as we embrace change in our response to our environment, as opposed to clinging to existing models. As Robert Kegan sees this process, it is a 'conversation' between us and the world that goes on throughout our lives:

> *This conversation is not one of continuous augmentation, but is marked by periods of dynamic stability or balance followed by periods of instability and qualitatively new balance. These periods of dynamic balance amount to a kind of evolutionary truce: further assimilation and accommodation will go on in the context of the established relationship struck between the organism and the world.*¹¹

So, clearly, we do not stop learning or responding to the right hemisphere when we are stably established in a one particular stage. However, we do use a relatively stable set of models and assumptions as the basis of our learning. We need models and assumptions to make judgements and act in the world. However, after a given period our models and assumptions become too constricting, like an old snakeskin, and we have to go through a period of vulnerability as we slough off the previous ones and grow our capacity to use new ones. Another kind of parallel to this is that of paradigm shifts in science as discussed by Thomas Kuhn.¹² Just as most science is 'normal science', working within a particular paradigm, but at times that paradigm has to be painfully outgrown, most of our everyday judgement is 'normal living', to only occasionally be disrupted by profound shifts in the assumptions on which that normality is based.

Even if you disagree with the theorizations of a particular stage theory, then, the larger point is that stage theories need to be organized in this way to be helpful to us. If stage theories are provisional but based on such a system, they can always be amended to take into account new considerations. In the remainder of this section, I will be drawing primarily on the stage theories of Robert Kegan,¹³ because they successfully integrate these different perspectives, as

9   II.5.g (pp. 206–7).
10 II.5.b.
11 Kegan (1982) p. 44.
12 Kuhn (1996).
13 Kegan (1982), further applied in Kegan (1994).

well as having a helpful account of tipping points that is compatible with Middle Way Philosophy. I do so in the provisionality of expectation that Kegan's theories could probably still be improved.

However, the problems with stage theories have multiplied beyond these integrative functions. I can identify three particular pitfalls that seem to attend them. One of these is excessive speculation and elaboration. Another is the confusion of stages with particular ideological beliefs. A third is the way in which stage theories lend themselves to abuse in the service of power through rigid hierarchization. None of these, in my view, amounts to a sufficient reason to dismiss stage theories, as their helpfulness outweighs these potential pitfalls. However, they do provide caveats that it is essential to be aware of before one makes use of stage theories.

The problem of speculative elaboration seems to have particularly attended the strand of stage theory dubbed 'Emergent Cyclical Levels of Existence', or 'Spiral Dynamics'. This began with the diligent research of Clare W. Graves,[14] and was continued by Don Beck[15] and Ken Wilber.[16] This approach does not only attempt to identify universal stages of human development, but also to stage cultural contexts in societies. It is one thing to take note of the *enabling relationship* that every cultural context has with the stages that people reach within it, but another to try to classify the cultural context more completely in relation to its psychological development. Sometimes, in the second half of this book, I will attempt to do the former – for instance, remarking in 3.a that some pre-Columbian Native American cultures allow an ideological level of judgement to a greater extent than we may recognize. However, that is very different from claiming that everyone in pre-Columbian Native American culture used this level of judgement, or even that all people in a given ethnic group were capable of doing so. Nor does it necessarily imply that some people in that group could not exceed that level in perhaps unexpected ways. The relationships between individuals and cultures are very complex, and thus very difficult to capture in a stage theory that is clearly justified by evidence. The most we can remark is that there seems to be evidence of some individuals reaching a given stage in a given culture, but this is only to

14 Graves (1970).
15 Beck & Cowan (1996).
16 Wilber's work is voluminous. Wilber (2005) is a helpful introduction to AQAL, which is an elaboration of stage theories.

classify the leading edge of the culture at best, not the overall psychological development of the culture as a whole.

These kinds of assumptions about staging and culture can also be associated with a second kind of pitfall, namely the identification of stages with particular ideologies or philosophies. The Spiral Dynamics model, for instance, separates out 'blue', 'orange', and 'green' stages[17] in a way that is primarily dependent on the differing ideological and value approaches in traditional religion, utilitarian application of science, and relativistic liberalism respectively. As we will see in more detail below (2.e and f), however, Kegan's approach, prioritizing the psychology of the conditions affecting judgement, instead offers two stages (interpersonal and ideological) that cut across these three, and are much more firmly based on the evidence of individual development. Kegan's approach here is much more compatible with one of the key insights of the Middle Way – that there is no basis for preference between opposing dogmatic ideologies, and one can flip from one to the other in an order that is not predetermined (for example, one could be a relativist first, then a scientific materialist, then a religious fundamentalist last, all resulting from a series of flips at the same level of psychological development). The Spiral Dynamics model appears to confuse the contingencies of history as they have predominantly appeared to us with necessary psychological processes.

A third potential pitfall is that of hierarchization and its abuse. This occurs when stage theories are used to classify people into hierarchies, rather than to merely to try to explain some aspects of their judgements at a particular point. Persons are complex, and their integration is asymmetrical.[18] The creation of a rigid hierarchy, then, cannot be justified by a stage theory, because stage theories do not give us a comprehensive enough basis for classifying persons in this way. Stage theories, for instance, miss out other aspects of the judgements of persons, of the kind that there might be some attempt to capture, say, in character typology, or in psychoanalysis of an individual's past experiences. All of these are partial tools, which put together may give us helpful information, but should not be absolutized. Hierarchies are often created for practical purposes, for instance in the chain of command of an army, but the value of such socio-politically necessary hierarchies should never be confused

17 Beck & Cowan (1996).
18 II.5.g.

with essentialized hierarchies of persons. Furthermore, even if we have practical reasons for creating a hierarchy and use stage theory to help us do so, this does not justify any absolute uses of power down the hierarchy. The use of power needs to be *practically* led, as I shall argue in the more detailed political philosophy to be included in a later volume of this series.[19] Even if stage theories provide us with helpful information that helps us to make our practical judgements, they cannot justifiably provide an essentialized shortcut way of making those judgements.

Not using stage theory for purposes of hierarchization or power has personal as well as political implications. It means, I think, that we need to observe a principle of reticence as well as a restraint of judgement. We need to remain provisional in all our judgements about what stage another individual (or ourselves) may have reached, which means that the appropriate use of stage theory is inseparable from the practice of the rest of the Middle Way. Even if we reach private and provisional judgements about this for practical purposes, it is important to keep them private, focusing on people's *judgements* at particular times rather than on their general status as a person. A principle of reticence about stage achievement, then, means that one should never voice an unconditional opinion either about someone else's stage, or about one's own. Making explicit or implicit claims about one's own stage of psychological development can be particularly damaging, as it is extremely likely to be perceived by others as a mere attempt to gain power by constructing social status. At best, it may be helpful to acknowledge obvious features of personal *judgement* (for instance, it would be difficult for me to write this book without being able to make at least some inter-individual judgements), but stop short of unconditionally identifying a person as a whole with a given stage.

Despite these caveats, however, my case is that the benefits of stage theory as a tool of reflection on the practice of the Middle Way outweigh the possible disadvantages, as long as we continue to consider it within the wider framework of that practice. In the remainder of this section I will be following the story of an individual's developmental life, from birth to a hoped-for wise old age, following Robert Kegan's categorization of stages. In the process we will be able to explore an important aspect of our systemic history, and one of immediate importance to all of us.

19 IX.1–3.

2.b. Birth and the Incorporative Stage

> *Summary*
>
> Following birth, we gradually differentiate the meaning of objects in our environment by interacting with them, in the process letting go of the undifferentiated meaning of the womb. We reach a tipping point of independence when meaning becomes predominantly differentiated. Negotiating that stressful transition from incorporative to impulsive stage requires secure attachment, which later provides a model for potential integration, and an implicit Middle Way balance to embrace our independence sustainably.

We cannot integrate what has not first been differentiated, and what is differentiated must first begin in a pre-integral unity (as already mentioned in 1.j). This pre-integral unity, in the history of each individual human life, is that of infancy, developing from what we may feel to be the undifferentiated bliss of the womb. That pre-integral unity provides the basis of *meaning* that makes integration subsequently possible as an option for us. In this chapter I will be drawing on embodied meaning theory and attachment theory, as well as psychological stage theory, to offer what I hope is a helpful account of the ways that the basic potential for integration (and hence for the practice of the Middle Way) is laid down in infancy. In my view, the distinction between meaning and belief is central for understanding this. That distinction also helps us avoid the pre-trans fallacy (see 1.j) whilst also recognizing the extent of infancy's conditioning effects. We need the right resources of meaning available to us from infancy to subsequently develop integrated desires and beliefs, and this dependency is often obscured by language that blurs the meaning-belief distinction, such as talk of 'meaning-making' (which is often actually 'belief-making') or 'sense of self'.

Robert Kegan defines our 'Stage 0' as the 'Incorporative Stage' from birth through to approximately 2 years of age. During this period, the infant is embedded (or 'incorporated') in a close embodied relationship with the mother (or other primary caregiver – I will refer to 'mother' for short). This relationship only starts to become challenged as the infant develops more physical independence as a toddler.[1] A tipping point is reached as the toddler becomes a self-

---

1   Kegan (1982) ch. 4.

sufficient child able to make judgements according to his or her own independent impulses, rather than constantly reverting to the security of the held relationship with the parent. The difficulties of this tipping point are dramatized in toddler tantrums, in which the desire for independence conflicts with parental responsibility.

If we understand this stage in terms of embodied meaning, we can also see it as the one in which initially undifferentiated meaning begins a slow process of differentiation, to eventually reach a tipping point where that meaning becomes more differentiated than undifferentiated. Undifferentiated meaning is an overwhelmingly 'emotional' source of meaning from which more 'cognitive' forms are able to gradually emerge. It is found in the initial bodily experience of the infant in the womb, and then when held in security by the mother. Basic reassurance associated with this undifferentiated security can then be subsequently renewed through close physical contact and by mutual gaze, which activates mirror neurons.[2] This activation of mirror neurons provides our subsequent capacity for empathy, for imitation, and perhaps even for other kinds of attunement between brain regions within or beyond one brain.[3]

From this it seems safe to assume that the undifferentiated state, and our associations with it, provide us with our basic model for the meaningful possibility of integration. We cannot integrate unless we have some prior idea of what a more unified state would be like. Further support for this assumption can be found from the attachment theory of John Bowlby,[4] which has been justified by observations of different patterns of response when infants are separated from their mothers, even in different cultural contexts.[5] Bowlby distinguished secure attachment from insecure types of attachment, dependent on whether the distress of separation could be quickly adjusted to, or whether infants showed longer-term patterns either of anxiously clinging to the parent or prematurely separating from the parent. But what makes attachment 'secure'? It seems to me helpful to interpret that security as the emotional power of undifferentiated meaning. It is because we have that prior experience of undifferentiated meaning, sufficiently renewed as it gradually becomes differentiated in early infancy, that we maintain a suffi-

2   Johnson (2007) pp. 36–41.
3   McGilchrist (2021) p. 201.
4   Bowlby (1977).
5   Ainsworth et al. (2015).

cient set of weak neural connectors from which meaning can then subsequently reintegrate in later conditions of conflict. Meaning, once it has been laid down, is long-lasting, perhaps permanent, as long as it can be at least occasionally renewed, which explains why expressions of parental bonding can constantly renew the sense of security needed by the child for it to integrate in future.

The gradual differentiation of this previously undifferentiated meaning may begin in some respects prior to birth, as the foetus comes to recognize and respond to particular stimuli that thereby have a specific symbolic value for it, particularly the maternal voice.[6] However, it is after birth, in the early months of life, that the infant begins to interact with its environment, and thus begins to differentiate elements of that environment as separately meaningful. At first, this may be directly related to the mutual gaze between parent and child, as the child follows the gaze of the parent to an object in the attention of the parent, and starts to imitate that attention.[7] Later, the infant may begin to interact with objects under its own initiative, but these objects remain formatted in terms of the infant's relationship with its goals and experience – what have become known as *affordances* rather than independent objects.

It is in the zone of these affordances that the 'cognitive' differentiation of meaning starts to occur, as the infant begins to organize its recurring impressions and associate them with each other to form patterns. Daniel Stern identifies some of the ways that infants develop these.[8] One of them is 'cross-modality' or 'amodality': the making of neural links that associate the experience coming from different senses (for instance, that the noise coming from a toy comes from the same source as what is seen). These abilities are then increasingly developed in relation to action and the experience of time, so that an infant is able to track the path of the same ball as it moves. Another early basis of organization is that of 'vitality effects' or 'patterns of feeling': these are gestalt formations from our (right-hemisphere) neural assembling of the sense experience of living movement, bringing together our recognition of movement, time, force, space, and intention or directionality.[9] These kinds of effects not only become central to our experience of the meaning of words

6 Jardri et al. (2012).
7 Johnson (2007) pp. 37–8.
8 Stern (1985); also discussed by Johnson (2007) pp. 41–5.
9 Stern (2010) pp. 3ff.

or other symbolizations of living things and their actions, but are also central to our experience of the meaning of dance and music.[10] It is these associative connections, not a system for representing objects as metaphysical entities, that then allows infants to identify objects in ways that are increasingly independent of the conditions of encounter: for instance, whether objects are detachable from their context, solid, substantial, and having a particular size and shape.[11] The meaning of differentiated objects, people, and events thus develops out of undifferentiated meaning, not vice-versa.

In its emotional impact, this first stage transition is probably the most difficult we will go through. It is hardly surprising that toddlers have tantrums, given the immense difficulty of moving from an undifferentiated and entirely secure world to one where separations from the parent are matched by separations of objects from one another for the infant's new, independent purposes. At the same time, however, the infant remains highly dependent on the parent and under her control in most respects. The tantrums, then, are over-assertions, as the infant briefly absolutizes its new independence in a flip from absolutizing maternal security. Such flips, as I discussed in *Absolutization*, are a feature of the fragility that comes with absolutization.[12] In the case of the infant, they are made more likely by a lack of the resilience that comes from developing confidence. Confidence builds up from a weight of embodied experience, including the both the weight of specific experiences justifying our beliefs and the weight of background security. The more meaningful we find something, the more confident we tend to be in it – a tendency that is both advantageous and disadvantageous. Our confidence allows us to learn and adapt, but without provisionality it can also make us more likely to accept beliefs that are merely familiar.

The transition from the incorporative stage to the next – what Kegan calls the 'impulsive' stage – is facilitated not just by the balance of emerging judgements in the infant, but also those of the parent. Prolonged separation from the parent can particularly endanger the transition. As Bowlby and other researchers into attachment have documented, the parents' own secure attachment facilitates that of the infant, because it enables the parent to let go of the infant when

10   Johnson (2007) p. 45.
11   Gibson & Pick (2000); Johnson (2007) pp. 47–9.
12   I.2.c.

it needs to assert its independence.[13] The parent, however, is faced with the Middle Way in the need to avoid both the absolute beliefs involved in hanging on, and those of letting go prematurely. Kegan summarizes the 'letting go' aspect of the parents' behaviour, allowing the child to 'emerge from embeddedness', as 'Does not meet child's every need, stops nursing, reduces carrying, acknowledges displays of independence and wilful refusal'.[14] On the other hand, the parent also needs to maintain continuity for the child, so that secure meaning remains in place as a basis of exploration. This continuity is maintained, Kegan suggests, through a shift from direct holding by the parent to embeddedness in the family unit.[15]

We thus have here the first emergence of the Middle Way, though in a form that remains highly implicit for the infant, even if it may be explicit for the parent. As we will see in relation to all the stages of human development, this emergence occurs in *transition*. Absolutizing beliefs are associated both with the old stage and with the new stage, with hanging on and with letting go. If we absolutize hanging on, we never pass the difficult tipping point that is required to develop greater adequacy in our responses. On the other hand, if we absolutize letting go, we lose the continuity and confidence that is required to engage effectively with the new stage, exhibiting what Bowlby called 'compulsive self-reliance',[16] interfering with our confidence both in forming relationships and in integrating our responses to the world. For the balancing feedback loop rather than the reinforcing one, we need to manageably import new information to modify our existing assumptions, not to wholly abandon those assumptions. It is only if we can manage the implicit Middle Way of this transition five times (as I will show in the rest of this section), to develop from zero stage to fifth stage, that we can then reach a point for using the Middle Way explicitly and consistently, rather than in a context where many other beliefs remain implicitly absolute.

At this stage, too, we have the emergence of the role of archetypal symbols as sources of inspiration to facilitate transition between stages, in the *transitional object*. The term 'transitional object' was first used by Donald Winnicott to refer to the role of teddies, other

13 Bowlby (1977); Ainsworth et al. (2015).
14 Kegan (1982) p. 118.
15 Ibid.
16 Bowlby (1977) p. 207.

soft toys, favourite blankets etc. as symbolic focuses of infant security.[17] These objects clearly act as symbolic prompts to the infant to evoke the security with which they have become associated, and can thus enable reassuring continuity in the infant's experience which is all about the persistence of undifferentiated meaning and acceptance, not 'illusions' or 'realities'. In this sense they fulfil precisely the role of archetypal symbols that I have argued for in *Archetypes in Religion and Beyond*.[18] Archetypal symbols are prompts aiding us to cope with the stresses of a changing environment by reminding us of a bigger context, so we can integrate over time. Again, this source of inspiration can be seen as an implicit aspect of the Middle Way, taking a form that can become more explicit later.

In some ways here we are already seeing that (*roughly*) 'ontogeny recapitulates phylogeny' as biologists put it: that is, that the development of individual lives in some respects follows the evolutionary path. The *implicit* development of the Middle Way as an option for infants as young as 1 or 2 years also reflects some *implicit* ways it could be used by animals with more developed neocortices, such as other primates, dolphins, or dogs. The development of increasingly differentiated meaning is certainly something we might expect to be shared by such animals, despite the limitations in the ways that such differentiated meaning then becomes associated with 'representational', grammatically propositional language. It is only between the ages of approximately 2 and 3 that children generally begin to string words together into sentences (in the 'telegraphic' and 'multi-word' stages),[19] which begins to apply that differentiated meaning in ways that we do not have evidence of other animals normally doing. The next, 'impulsive' stage, as we will see, begins to involve more specifically human contexts and problems.

---

17 Winnicott (1953) – to credit the use of term only.
18 Ellis (2022).
19 De Villiers & De Villiers (1978), ch. 3.

2.c. The Impulsive Stage

> *Summary*
>
> In the impulsive stage, the child has a viewpoint, but is subject to unintegrated desires ('impulses') and still lacks the ability to integrate these by recognizing alternative viewpoints, whether internally or externally. He/she remains dependent on the family, but begins to transition to the next stage in contact with the wider community, as the child begins to develop a coherent role in that community along with other consistent models of the world. Finding the Middle Way for transition means making use of rules and authorities that at this stage give context, though at later stages they might become absolutized.

The term 'impulsive' is given by Robert Kegan to what he treats as 'Stage 1' of human development (the incorporative stage having been 'Stage 0'). It refers to the period roughly between the ages of 2 and 7, during which the child is 'embedded in impulse and perception'.[1] I take this to refer to the child's own impulses, in contrast to the state of being held or 'incorporated' by the parent in Stage 0. These impulses are contrasted with the 'enduring dispositions and needs' that Kegan finds to be more typical of older children, in which impulses are contextualized at another level of complexity. This stage is referred to by Erikson as the stage of 'initiative v guilt', and by Kohlberg as 'punishment and obedience orientation',[2] suggesting the way that a focus on unintegrated desires at this stage also implies a relatively narrow focus on responses from others that can enable or frustrate those desires. It is also referred to by Piaget as 'pre-operational', focusing on the lack of an 'operational' use of reasoning to form a consistent model of the world.

As mentioned in the previous chapter, it is very much the development of propositional language that helps to set off this stage. The infant was gradually collecting meaningful impressions of the world, but the child can now form and regularize specific *beliefs* about the world using sentences that relate subjects, objects, and verbs in an assumed state of affairs. One might then assume that a child can thus assemble *coherent* beliefs about the world, but as yet it lacks many of the tools required to create this coherence.

---

1   Kegan (1982) p. 135.
2   Ibid. p. 134.

Instead, inconsistency of belief and experimentation with meaning are much more characteristic of this stage, which we might also call the 'fantasy stage'. Beliefs are instead much more in the immediate service of impulse, and as a result commonly deluded – though in a way that can be much more readily corrected than the delusions of mental illness in an adult, because they are also accompanied by constant enquiry. Kegan aptly summarizes many of the differences in both the use of language and the worldview of Stages 1 and 2:

> *The younger child uses language as an appendage or companion to her means of self-presentation and social intercourse; for the older child, language is the very medium of interaction, central to the social presentation of the self. The younger child's life is filed with fantasy... (being Spiderman); the older child has taken an interest in things as they are and fantasy life is about things that actually could be (being a doctor). The younger child will engage her parents in the middle of a conversation that she has already started on her own, as if she has trouble keeping track of which portion of a conversation she has actually had with you and which she has only imagined with you, or as if she takes it for granted that her private thinking is as public and monitored as her spoken thinking; the older child never does this....*[3]

The distinction between the first and second stages can also be understood in terms of a flexibility of viewpoint. The younger child is able to understand and describe the world from his or her viewpoint, but this viewpoint remains fragmented precisely because he/she cannot yet recognize other viewpoints, whether these are the viewpoints of others, or his/her own viewpoints at other times. This has been misleadingly described as 'theory of mind', though it has nothing to do with theories of mind. Laszlo et al., in a discussion of cognitive maps, have a striking example:

> *A child can be asked, for example, to describe a table-top landscape that includes several farm animals, a small lake, and a papier-mache mountain, all as seen from the eyes of a doll placed nearby on the table. He will do a passable job of this, if he is near the doll and his own perspective is also that of the doll. When asked to walk to the other side of the table, however, and again to describe the landscape as seen by the doll, which has not been moved, his description changes dramatically to his own physical perspective, and utterly disregards the view from the side of the table where the doll remains, and where he was in fact just standing. The child is said to be* **perceptually bound.**[4]

---

3   Ibid. p. 136.
4   Laszlo et al. (1996).

In this and similar tests, younger children give the same limited answers as those with autism or with right-hemisphere damage, suggesting that the ability to connect together perspectives beyond one's current one in this way is very much a product of right-hemisphere neocortical development,[5] and is developed chiefly in relation to a social understanding of the viewpoint of others. Without the ability to put together these differing perspectives, the potential integration of children in the impulsive stage is obviously severely limited.

Such continuing limitations, in the context of a complex human society, obviously mean that a child of this age still needs a good deal of social protection and support. This support is typically offered by 'the family' – that is, an immediate supportive social group, whatever form that group takes in a particular cultural setting. However, as the child moves on from the impulsive stage, that family unit will need to be increasingly absorbed into a larger social sphere within which the child moves. The child will also need to see itself increasingly as an individual taking a role in that wider world – typically the world of school or of other communal contexts where children are welcomed. Though very much socially defined and controlled, that role will be the child's first initiation into individuality, because we need to shape ourselves into awareness initially using the template of others' expectations. Whilst still in the impulsive stage, the child lacks that individuality: not because we cannot see individual characteristics in the child, but because the child has not consistently shaped those characteristics for itself.

The transition to the following 'imperial' stage, then, is clearly one of integration made possible by new developments in neocortical capacity. The child moves from a world of inconsistent beliefs, organized for it by others, to one where consistent beliefs and purposes are at least an expectation. Although the learning process to get beyond this tipping point seems to be right-hemisphere based and to involve a balancing feedback loop of adjustment, once reached that consistent view of the world also becomes the domain of the left hemisphere. Not only the child's view of the world, but also its view of itself, becomes more consistent, as the child increasingly adopts a role – so much so that Kegan calls the Stage 1 to 2

---

[5] McGilchrist (2021) pp. 200–1.

transition 'the birth of a role'.[6] The expectation of that social role is often developed and reinforced through schooling.

Just as in the previous transition, the Middle Way is implicitly required to navigate the transition successfully, even if the child is still a very long way from grasping it explicitly. On the one hand, if the child identifies too obsessively with a new independence from the family (or is forced into premature independence by circumstances), a shutting down of learning potentialities is likely to result in the long term. On the other hand, if the child hangs onto the old context with its security of outlook (for instance, in cases of school phobia), he or she becomes thereby limited in engaging with the conditions of the world and the confidence that is required in dealing with them. The relationship between meaning and belief is brought to the fore here, as if we prematurely limit our fantasies to focus on the 'real world', we lose creativity in engaging with that world, but if we hang onto the fantasies, the world becomes increasingly likely to intrude disruptively. The same balancing judgements need to be continually made throughout our lives, even as adults – when it is often too easy to give up play or the development of meaning to focus only on those beliefs that we judge to be of immediate practical relevance. The more complex and unpredictable the environment we have to deal with, the less effective that strategy is likely to be in the long term.

Kegan gives a moving example of this transition from a conversation with his own daughter when she was around 6 years old.[7] The child had been playing across the street with a friend, but then returned with a new toy that she claimed to have found in the street: 'since it's lost and there's no way of finding out who it belongs to I guess I can just keep it.' However, she wanted to talk to her father about this and to justify herself, obviously betraying an unease. Kegan quickly worked out that she had stolen the toy from her friend, and wanted him 'to help her with the turmoil she was feeling over it'. He goes on

> *Had she been firmly embedded she would have had no trouble acting on her impulse.... She would not be directing my attention to how she got the toy and the legitimacy of her keeping it; she would just be playing with it. Had she been firmly embedded in the next balance, she would not be engaged in such a*

6   Kegan (1982) p. 137.
7   Ibid., pp. 146–7.

> *conversation either. She would be able to* have *the impulse... and experience herself as jealous, which might be enough to keep her from having to take it; or she would purposely and planfully steal it, and she would quite intentionally get herself upstairs to hide her booty away before I saw her.*

By interacting in this way with a psychologically aware father who realized what she was up to, however, she put herself in a situation where she could be encouraged past the tipping point, by her mixed motives being discussed explicitly and placed in the wider moral context of social rules against stealing and lying. Of course, if the parent had not been so integrated or aware himself, he might have implicitly accepted the unintegrated lie, or alternatively have imposed the moral rules on her without helping her to understand her feelings and motives, thus failing to use this opportunity to encourage her to develop.

Here we can see a phenomenon that will be constantly noteworthy as we continue to work our way up the stages: namely that a set of social assumptions that is integrative at one level can become dogmatic at the next. From the standpoint of the incorporative stage, the mere awareness of a parental prohibition interfering with the fulfilment of an impulse is a step forward. From the standpoint of the imperial stage, that prohibition becomes an internalized rule, a feature of the phenomenal world and the child's relationship to it. In the process, that rule offers a new tool for incrementalized objectivity of judgement. However, as we will see at the later stages, adherence to a fixed rule can itself become a source of reinforcing feedback loops and left-hemisphere based rigidity. The tools of development vary in the significance at each level, but the patterns of development nevertheless follow the systemic form of reinforcing and balancing feedback loops throughout.

2.d. The Imperial Stage

> *Summary*
>
> The imperial stage offers consistent beliefs about an immediate, concrete, assumed reality. Roles and rules are followed instrumentally, but without empathy towards others' viewpoints. A significant minority of adults are stuck at this point. Transition to the interpersonal stage requires imperial motives to begin to resolve conflict with others in a way that is not just instrumental, but has genuine loyalty. A new capacity for balancing feedback is developed through relationship.

We are now at the stage of the stamp-collecting 9-year-old: confident about the world, engaging with concrete things with consistency, and as yet untroubled by questions about personal relationships. Robert Kegan calls it 'imperial' because of the apparent self-sufficiency with which children of this stage (approximately 7 to 13) interact with the world. In practice, that world is still strongly shaped and circumscribed by the surrounding culture, but the imperial child is able to take that culture completely for granted, and pursue his or her goals within it as though it was entirely fixed. That suggests that this stage is the one where left-hemisphere dominance starts to become entrenched.

In Piaget's understanding, this is known as the 'concrete operational stage' – operational because of the use of consistent 'logical' models, concrete because applying only to experienced objects, not to abstractions or hypothetical constructions. Children at this stage can not only understand and use propositions, but begin to relate them to each other using reasoning, as long as they are related to immediate experience. For example, children learn the conservation of number and volume – that a given number of objects does not change if you merely rearrange them, and that a volume of water remains the same whatever shaped container you put it into. They also learn that equivalences of this kind can be reversed, and that numbers or amounts can be placed in a rank order.[1] At this stage, then, propositions are overwhelmingly being used to understand the world in direct relationship with experience, either through induction or through deductions that can be closely related back

---

1   Piaget (1954) ch. 1.

to experience. The absolutizations of metaphysics are not yet possible, but the tools that can later be used for this purpose are being developed.

Kohlberg talks of 'instrumental orientation' to describe the moral aspect of this stage. Children here become aware of the consistency of moral rules, just as they become aware of the consistency of 'facts', but are not thereby necessarily committed to them. Rather they are concerned with the impact of moral rules on themselves. The degree of integration involved in creating a consistent view of the world helps children to develop the suppression or repression necessary to follow rules, yet that same consistency may enable purposeful breaking of rules – of the kind Kegan's daughter (discussed in the previous chapter) was not yet quite capable of. Children of this age can be cunning, but usually for narrow ends. Before they become capable of more reflective rule-following or rule-breaking, people need to develop more intrinsic motivation through a developing capacity to care for others, as we see in the next, interpersonal, stage.

In relation to others, children of this age develop a *role*. As Kegan remarks, its significance is not so much that of any particular role (much specialization is rare at this stage), but more 'the organization and exercise of what a role is'.[2] That role allows increasingly self-conscious individuality and independence, for instance having one's own bedroom, or making one's own way to school. The role in relation to parents is that of child, in school that of pupil, in other groups or organizations that of junior participant, and so on. The child becomes aware that social rules vary in different roles, and that different types of behaviour are expected in different places. These social rules also develop scope for competitiveness, especially in peer groups. Adults are likely to reward independence in the performance of these roles and fulfilment of tasks in them, so that imperial children compete with each other to excel in them, without questioning their framing. Turning a lesson into a competitive game thus motivates imperial children, and often also the imperial tendencies within those who have reached further stages.

What this stage lacks, in comparison to the interpersonal stage that follows it, is a basis of social values arising from sensitivity to others and their perspective. The imperial child is thus 'unsocialized' or

---

2 Kegan (1982) p. 162.

'immature' when seen in terms of the expectations we are likely to have for an adolescent or adult. In particular, the imperial child may lack *internalized* sources of motivation for acting in the ways that others would prefer when this clashes with their short-term desires – what is often thought of as a conscience. Kegan illustrates the gap with the story of Matty, a 16-year-old who arrives back home from a party at 2.30 am when he was meant to be in by midnight, ready to conceal the fact or to make excuses.[3] Matty understands that his parents have a distinct point of view from his own, but is not genuinely concerned about that point of view – only with avoiding negative consequences to himself. What he has not yet done is to integrate his parents' point of view with his own to create a wider contextualized view. As a teenager, however, his parents expect him to have done so – to have adopted the basic values of a responsible citizen rather than only the independence of an older child. He is 'in over his head'.

According to a range of studies considered by Kegan, a small but not inconsiderable proportion of adults never manage the transition beyond the imperial stage, with somewhere between 5 and 13% entirely limited to it, and a further 8–23% still partially using it as a basis of judgement.[4] It seems likely that this proportion includes people that we may describe as having 'learning difficulties' of one kind or another. It may well also include those often described as 'sociopathic' or 'narcissistic', including some people who reach positions of power, precisely because they are not concerned with any consistent constructed view of what others think of them, only with the short-term feedback. In a review of the research on narcissistic leadership, Seth Rosenthal writes 'Narcissists are thought to seek leadership positions specifically to garner the power that enables them to "structure an external world" that supports their grandiose needs and visions.'[5] This is typical of the imperial level of operating, with its fantasy play in which one's own world is shaped without the intrusion of wider social concern of any kind.

The arrested imperial stage in adulthood needs to be distinguished from that of the later ideological stage (2.f) in which people follow rationally rigorous beliefs that they sincerely believe in and take responsibility for. Think of the distinction between the typical

---

3   Kegan (1994) pp. 15–28.
4   Ibid. p. 194.
5   Rosenthal (2006).

judgements of a sincere but inflexible ideologue – say Pope Benedict XVI, with those of a childlike leader apparently unaware of his own inconsistencies, like Donald Trump. At the ideological level people can at least change their positions within the terms of the ideology, rather than merely adopting superficial features of the ideology for short-term goal-fulfilment. However, the route to such wider consistency of values lies through the interpersonal stage, rather than only in the development of consistent concrete views of the world that we get in the imperial stage.

The transition to the interpersonal otherwise generally occurs in adolescence, through integration of the conflicts with others created by the imperial approach. Obviously, cultural expectations, the biological conditions of the maturing body, and the development of conceptual capacities would also play a necessary role here. The development of greater empathic capacities (especially in girls, who reach sexual maturity faster than boys) may open the key route through awareness of others' feelings, which then triggers the need to overcome conflicts with those feelings. In other cases, adolescents may need a strong push from cultural expectation to start adopting a more socially responsible approach. These expectations may come from parents, from school, or from other adults. However, these expectations alone may just have the effect of producing entrenched conflict if the young person does not find their own independent path to an interpersonal level of relationship.

In most cases that path seems to be found in continuity with imperial motives through developing mutual relationships with peers – most often schoolmates, but perhaps also workmates or other friends. The more ambitious the enterprise, the greater the likelihood that imperial competition will need to be supplemented by cooperation, and that a bond of mutual trust will need to develop to make that cooperation possible. My own personal experience of early adolescence clearly followed that pattern, as I became involved in a group of friends involved in a very typical (male) imperial competitive activity – wargaming. However, the social arrangements this required, and the increasingly creative development of new games or of other spinoff interests, also required the gradual development of bonds of friendship and increasing intimacy. Kegan gives the example of a boy, Richard, who was put on a special programme in late adolescence because he was considered 'unemployable' – and at first he was very unreliable and showed no respect for contracts or working relationships. However, the challenging focus of the working tasks he was engaged in – building boats – attracted his imperial interest, and soon required more responsible attention to his relationships in addition to the direct concrete actions in the work. Within a year of such work, his attitude was transformed as a new maturity emerged.[6]

The building of bonds that enable one to begin working effectively in relationships of trust to others may also be culturally supported by the communities of practice discussed by Alasdair MacIntyre as the key context for the development of virtue.[7] The practice of football as a source of virtue, for instance, does not only consist in the skills of kicking a ball accurately (which may well be acquired at the imperial stage or even earlier), but also the acceptance of the values of a team ethos in live relationship to others with shared goals. These shared values provide a kind of template of social expectations in relation to which a young person can develop empathic awareness of others' needs and expectations in relationships of increasing mutual trust. Many virtues are social, and seem to have their crucial point of development in the transition from imperial to interpersonal stages.

In transition from the imperial to the interpersonal stage, then, we clearly develop a basic capacity to resolve social conflict through the exercise of empathy, together with an extension of our capacity for consistent thinking so as to include both explicit and implicit contracts with others. As in previous transitions, this seems to require the exercise of right-hemisphere functions to make the transition possible through the necessary balancing feedback loops, even though both hemispheres are needed in both stages. The tendency to get stuck, making judgements that result in reinforcing feedback loops when there is a possibility of transitional judgements, is similarly likely to be due to absolutization in which the left hemisphere is over-dominant.

It is at this point of transition to the interpersonal that we enter the community – a point often traditionally marked by symbolic rites of passage into adulthood. These rites of passage often include periods of separation, transition, and re-incorporation into society, symbolically replaying the individualism of the imperial stage

6   Kegan (1982) pp. 173–4 & 181–2.
7   MacIntyre (1981); II.5.g.

followed by the wider acceptance of the interpersonal stage.[8] The virtue of fidelity seems to be the one centrally tested both in symbolic rites of passage and in the longer-term process of stage transition.[9] This fidelity will be required in the adult human roles not only of work, but also crucially of marriage and child-rearing. We will see more in the next chapter of both the advantages and the limitations of that disposition for the long-term integration of human individuals and societies.

8   Markstrom et al. (1998) pp. 338–9.
9   Ibid. pp. 345–6.

2.e. The Interpersonal Stage

> *Summary*
>
> The interpersonal stage involves thinking based on assumptions that are consistently in the service of the group, maintained through socio-political value foundations of authority, loyalty, and purity. Wider values of justice, freedom, and care are interpreted in terms favourable to the group. Transition to the ideological stage requires consistently more universal expectations from an alternative group, such as at university, with the Middle Way needed to navigate it.

The interpersonal stage is in most respects the beginning of adulthood, and it marks the developmental stage of the majority of adults. It plunges us into a world of ties and responsibilities: the social context where others need to rely on us taking a particular role, whether that of a spouse, a parent, an employee, a friend, a soldier, a tradesman, or even a leader. These roles are far more effectively fulfilled if we perform them with intrinsic motivation – if we have a sense of duty developed from our feelings for others, rather than just a fear of personal consequences. This stage, then, activates our capacity for solidarity with others and for acting as a reliable member of society. At the same time, it still incorporates the capacities of the impulsive and imperial stages. We do not have to lose our ('impulsive') capacity for imagination, and for considering different possibilities, and we clearly still need our ('imperial') capacity to understand and manipulate the world consistently, to fulfil our role in this new interpersonal world: so this stage, like all new stages, incorporates all the capacities of the previous ones.

For Piaget, this period marks the beginning of the formal operational stage, in which a capacity for merely abstract and hypothetical deductions begins to develop. This becomes important for the understanding and application of shared group understandings of the world and rules of behaviour. The shared view of the world accepted by a group is largely not a result of individual learning in which theory is related to experience, but rather of abstract logical relationships between propositions. This can be seen most starkly in scenarios such as that of war. In the Russo-Ukrainian War going on at the time of writing, the two sides not only have different moral priorities, but entirely different definitional accounts of the facts of the situation, such as what is the legitimate territory of each side, who is a 'Fascist', or what counts as a 'military target'. These two different socially-based accounts are of course shared in all forms of communication, including mass and social media. To question any one element of this socially-based account implies questioning of them all. To a lesser extent these interrelated group-based propositional beliefs are also typical of less seriously fragmented and more partially opposed groups, such as opposing political parties, classes, professional groups, or groups created by educational experience or strong economic interest. From roughly the age of 12 we begin to be capable of adopting these abstract beliefs in a way that takes our responsibility to do so as an aspect of mutual social duty. To believe the other side's lies would be to betray one's comrades, through the capacity for comradeship that we gained in adolescence.

To adopt this group-defined worldview does not, of course, mean that we necessarily *absolutize* its beliefs, but it is much harder to avoid that absolutization if we do not have a wider context in which to place the group's beliefs. That context only arises when we become aware of potential conflicts between our own group's view, and the views of another group that we are obliged to start taking seriously – a conflict that is likely to start occurring in preparation for the Stage 3 to 4 transition. Adolescents are thus at a peak stage of vulnerability to indoctrination or cult-recruitment: able to understand the logically connecting assumptions of absolute belief systems, but often not able to contextualize them, unless another person offers an immediate contrasting perspective that also carries interpersonal weight. That many adults stay at this stage of vulnerability for the rest of their lives also helps to explain how easily religious groups and populist political movements gain and maintain their adherents.

Just as the imperial child learned a new objectivity in creating a consistent view of the concrete world, the interpersonal stage also demands a new objectivity: that of consistent awareness of others and their emotional states and beliefs. In the light of that awareness of others, we also experience both the advantages and disadvantages of the dominant group view of the world held by the significant others that we become sensitized to. Moving independently in a wider society, the adolescent is no longer confined to the overriding influence of his or her parents as the focus of this sensitivity. Peer-bonding, both for sexual relationships and wider comradeship, is an obvious feature of adolescence. In modern Western culture, that peer-bonding can be culturally separated from the parents sufficiently to enable the formation of new challenging values, that may in some respects be in conflict with those of the parents: counter-culture, the teenage music scene, the latest online crazes. In traditional societies, however, peers are embedded in the same culture. Whether the movement is pro-cultural or counter-cultural, though, the adolescent gains the support and confidence of the group, potentially along with unquestioned subordination to its values.

I have found considerable light shed on the values of the interpersonal stage by the work of Jonathan Haidt, who has used empirical research in social psychology to develop a theory of foundational values.[1] These values help to make sense of dominant cultural norms and of political conflicts in society, but Haidt does not seem to have made what for me is the increasingly obvious connection with psychological stage theory. Haidt identifies six core types of value used in social and political judgement: loyalty, authority, purity, justice, care, and freedom. He does not distinguish these from *moral* values, but I think it important to do so: thus I will refer to them as sociopolitical value foundations, not 'moral foundations', which is his term. Socio-political value foundations tell us about the raw material we can then morally shape, not about morality as a whole, but nevertheless this raw material is very important to understand. It is raw material that emerges in the course of human psychological development.

Of these six types of value, he finds three that in explicit socio-political judgements are only associated with traditional societies and with conservative groupings in modern society: these are loyalty, authority, and purity. These are all notably group-facing, interpersonal values. Loyalty prioritizes our specific bonds with others. Authority prioritizes the values emphasized by a specific person or persons who have been widely accepted by a group, from gang leaders to teachers to presidents. Authority may also lie in the traditions of the group itself, such as the performance of a ritual that has been done the same way for generations. Purity tries to maintain the integrity of things valued by the group, rejecting anything that might pollute or compromise those things – whether these are, for instance, rituals, lands, or institutions. These are also notably

[1] Haidt (2012).

the values of the interpersonal stage: loyalty to individuals of the peer group, acceptance of the authority adopted by that group, and defence of the core values of that group against threats from beyond it.

Haidt also found that traditional and conservative groupings appreciated the other values: justice, care, and freedom. Justice and care, particularly, might seem like highly interpersonal values, concerned with the welfare of others as well as ourselves. However, the difference between traditional and modern (or conservative and liberal) sets of values lies in how these other values are interpreted, and whether they entirely overrule the more relative values of the group. Those at the interpersonal stage are much more likely to value justice *for the people in their group* and care *of the people in their group*. It is the universality of these values that requires the Stage 4 'institutional' or 'ideological' thinking much more typical of liberal approaches in the modern context. Adolescents also clearly value freedom, but *their* freedom or that of their peers, not somebody else's on the other side of the world.

Interpersonal values, then, are relative values. That is not because those thinking in interpersonal terms are explicitly relativists, but rather that universality does not occur to them as a practical basis of judgement sufficient to be prioritized over more particular, group-based, values. Those thinking interpersonally may indeed use language that in principle is universal – for instance, the language of Christianity – but in practice interpret it in a way that only prioritizes the near neighbour, not the big picture. From the standpoint of the later stages, then, interpersonal thinking can seem hypocritical: for instance, 'universal love' can mean cutting welfare support or restricting immigration. For the person thinking interpersonally, though, the concept of 'universal love' has an archetypal value that reminds them of their interpersonal values,[2] rather than being a prompt for the type of reasoning that identifies inconsistencies. To require consistent ideological thinking of them may be to ask for a valuing of abstract consistency that is not yet meaningful to them – to put them 'in over their heads' in a similar fashion to the adolescent still in the imperial stage asked to regularly turn up on time for work.

2   Ellis (2022) 1.h.

The studies drawn on by Kegan found 14% of the general adult population operating solely at the interpersonal stage. However, when one also includes those working inconsistently between the imperial and interpersonal stages, this rises to 22%, and when one also includes the large number working inconsistently between interpersonal and ideological stages (32%), this rises to a total of 54%.[3] This tells us much about the popular base for socio-political decision-making: that is, that the majority of the population (at least in Western countries – Kegan's studies come from the US) are often likely to prioritize relative interpersonal values – and thus reject universal or liberal values that are not sufficiently meaningful to them. This does not prevent interpersonal thinkers from supporting universal causes, but they will only do so in a way that is motivated by loyalty to their group. In this way, past working-class support for the universal ideological thinking typical of socialism is easily explicable – because working people who thought interpersonally were voting for their class as part of a class struggle. However, when that interpersonal framing is replaced by the more universal framing of the next stage, working-class people who are at the interpersonal stage may well abandon socialist approaches. It appears alienatingly impersonal to them, or pitches them 'in over their heads'. In a study matching socio-political value foundations to voting behaviour in the 2012 US presidential election, foundations of purity, authority, and loyalty turned out to be a far better predictor of Republican voting than any demographic indicator, whilst stronger foundations of fairness, freedom, or care (particularly fairness) predicted Democrat voting.[4]

The transition from the interpersonal stage to the next one (which I will call the 'ideological stage', though Kegan calls it the 'institutional stage') is related, like the previous stage transitions, both to developments in 'cognitive' capacity and to the new emotional demands of different kinds of social interaction. Where the interpersonal stage begins what Piaget calls 'formal operations' – namely the consistent use of reasoning – these operations are limited at the interpersonal stage by the horizons of the group and its assumptions. It is one thing to use the connections between different beliefs to justify (or rationalize) the beliefs of the group, but another to

3   Kegan (1994) p. 195.
4   Franks & Scherr (2015).

identify inconsistencies that lead one to question group norms and adopt wider alternatives.

To start using reasoning in that way just as much requires a wider emotional contextualization of the group and its concerns in wider conditions: not that one's group becomes unimportant, but that it takes its place in a wider emotional setting of universal concerns. Those might be concerns about justice for people far away, or for animals previously excluded from sympathetic consideration. They may well involve an identification with a universal concept, whether that is human rights, obedience to the will of God, the coming revolution, adaptation to market forces, or the value of artistic expression for the individual. This is unlikely to happen only through the encounter with abstract ideas by itself, but more likely through identification with an encountered group with such universal ideological beliefs. A group of politically- or religiously-motivated friends, an influential lecturer, or the demands of a bureaucratic organization are all common examples of vectors catalysing the tipping point between interpersonal and ideological ways of thinking. Initially interpersonal motives in relation to a universalizing group come into conflict with those of older, more traditional groups, and this conflict is resolved by making the beliefs of the older group subordinate to the wider ideological vision of the newer group.

This is evidently why Kegan identifies the key bridge between interpersonal and ideological stages as 'a time-limited participation in institutional life' that offers 'opportunities for provisional identity that both leave the interpersonalist context behind and preserve it, intact, for return'.[5] He mentions employment and military service as possible institutions, but perhaps the most common in the West now is college or university study. Employment or the military are likely to be rigorous in forcing us to obey a new wider set of imperatives, but their heavy reliance on extrinsic motivation may also make them ineffective as media for discovering wider universal values. Higher education, on the other hand, perhaps has as its main implicit social purpose the facilitation of Stage 3 to 4 transition. Here we may learn not only that our parochial home life did not offer all the answers, but why. Of course, the effectiveness with which higher education performs that role can also be questioned, and doubtless varies between courses and institutions (especially

---

5   Kegan (1982) p. 191.

between countries, where higher education culture varies a great deal). However, even rather narrowly focused vocational courses (such as engineering or management) may require the development of critical thinking skills that may at first be applied only quite technically – but are then available for application in more general ways.

Of course, Stage 3 to 4 transition may never be completed, even though helpful cultural conditions such as higher education have been provided, because it depends entirely on the individual's capacity to make judgements that gradually move over the tipping point from immediate social influence to a wider application of universal principles. To some, such principles may seem alien and deadening, like being confined to a classroom forced to rote-learn boring abstractions whilst the attractions of the 'real world' beckon in the sunlight outside the window. Others may be influenced by perceived social or economic advantage to fake Stage 4 thinking, or adopt it only in certain advantageous contexts but not others: hence the 32% in Kegan's research who moved inconsistently between Stages 3 and 4 thinking. Many people, for instance, are obliged to be systematic at work, but this degree of general thinking remains highly domain dependent, and is never applied at home, where unquestioned family power-relationships prevail. For those who fail to make this transition and are alienated from it, conservative socio-political attitudes provide a ready focus for projected conflict.

As in the previous transitions (but more poignantly here, because it is less reliably completed), a micro-Middle-Way is required to successfully develop into the ideological stage. If we continue to believe that the old thinking is complete, we will never make the transition, refusing all opportunities to do so. If, on the other hand, we absolutize the new thinking by embracing it prematurely as a whole, we may fail to maintain our motives, which are still primarily those of the interpersonal stage. Such is the case, for instance, in a young person from a fairly traditional background who develops a sudden strong emotional attachment to another person or a group strongly embedded in ideological thinking: they may run away and live with a much more mature person, or they may join a dominant political group with strong ideological motives. However, the most likely long-term outcome is disillusionment, as the new context begins to seem alienating and rootless, and a return to the old context where old ways of thinking will probably resume. Transition to the ideological stage is not just a left-hemisphere based intellectual process to be completed in the abstract, but also a transplantation from one type of group to another.

2.f. The Ideological Stage

> *Summary*
>
> The ideological stage reflects the adoption of universal models, though these can be used either to inspire or to exclude new examination. It allows individual self-sufficiency in relationships, principial ethics, progressive politics, and explicit metaphysical belief (both positive and negative). Sceptical argument, questioning of one's persona, or existential dissatisfaction can all help trigger transition to the fifth stage. However, the institutional entrenchment of ideology is likely to make that transition partial in most cases.

The fourth stage is characteristic of adults educated to graduate level, and/or engaged in professional responsibilities. It requires the consistent use of reasoning to develop beliefs that are universal in scope, but nevertheless dependent on particular models. I call it the 'ideological stage' because of that very consistency of belief, which is characteristic of political ideologies, as well as the consensus models behind religious traditions, business operation, socially-engaged professions, scientific research, and schools of artistic activity. We engage with that consistency of belief alongside others who share it, but at the same time need to challenge those with more parochial group-based models. Kegan calls it 'institutional' because of the ways that involvement with institutions tends to promote that rationalized consistency. However, I prefer to focus on the ensuing universal beliefs as the most characteristic feature of this stage, howsoever arrived at.

Universal beliefs are double-edged, as I discussed in *Absolutization*.[1] On the one hand they offer an *aspiration* to check our beliefs so far against any and all new examples that we encounter, so that they remain universal as far as we can tell: that is what I call *bottom-up* universality. On the other hand, universality can consist in a dogmatic assumption about the total application of a given belief to all possible examples, which is then applied so as to *preclude* examination of new examples: this is *top-down* universality. The bottom-up version offers a balancing feedback loop in which universality has an archetypal function as a concept, inspiring us to keep looking and adapting, but the top-down version is a

1 I.8.a.

reinforcing feedback loop that continues to repress all awareness of alternatives or exceptions.

The ideological stage can function for us in either of those two ways. On the one hand, we learn how to use universal inspiration to keep developing our objectivity in a provisional fashion. On the other, we settle into a rigid scientific and/or bureaucratized view of the world, in which everything needs to fit in with a grand purpose that itself is beyond question. Ideological thinkers and administrators have shaped the world in the image of their beliefs for many centuries, from the missionaries who shaped foreign societies in terms of their religious beliefs, to the political revolutionaries or statesmen who have insisted on their vision of a new society against all opposition, to the managerialists of public administration or business who try to fix everyone's awareness towards key objectives by endless checks and form-filling. These people have a vision of what they want, which at its worst lays waste to all evolved complex systems in their path, but at its best can allow further learning to interact with the complexity of systems and engage that vision with underlying conditions. To arrive at more provisional forms of universality, though, we have to first learn to universalize in the first place – the road to the fifth stage thus lies through the fourth.

In emotional life and relationships, the ideological stage represents a new self-sufficiency founded on reflective control of our own judgements as individuals. That doesn't mean that social judgements are necessarily rejected, but that these have a wider universal context for comparison. This offers people the independence required to make an unattached life emotionally sustainable, or a marriage much more of a contract between equals and less of a mere mutual dependency. That independence has allowed people to move away from their parents or other relatives, often giving priority to professional life to create a new mobility that follows economic opportunities. It can also be associated with serial monogamy as a way of life, as people maintain the confidence and independence of mind to end an existing relationship despite all its sunk costs, then negotiate a new one. Kegan gives the example of 'Michael', an unattached man in his 30s for whom no new relationship with a woman is ever quite right. He realizes he has 'high standards' that reflect the independence of the ideological stage, though he also begins to recognize the costs that these 'high standards' impose on him.[2]

2   Kegan (1982) pp. 221-5.

The ideological stage becomes intrinsically motivated in a way that the interpersonal is not, though one that is dependent on the consistency of 'principles' maintained internally. Morally speaking, then, the ideological adult is capable of acting on the basis of a sense of duty, even if this goes against immediate group interests – say, caring for a stranger, or refraining from taking revenge despite injuries to his or her group. Although this sense of duty can in its turn be rigidified, it offers a moral stretch from mere social morality, and it may be expressed by any of the three main forms of normative ethics – consequentialist, deontological, or virtue ethics.[3] We may question social conventions by prioritizing actions that have better outcomes for all – for instance, refusing to participate in traditional female genital mutilation. We may question social conventions by appealing to more general abstract principles, howsoever derived – for instance, refusing to lie even when there are strong immediate social incentives to do so. We may question social conventions because of a longer-term commitment to our own development or that of others – for instance, going on a retreat despite the opposition of one's immediate family. The recognition of this universalizability of ethical thinking begins with Kant, who identified it solely with deontological (principial) types of ethics. However, any type of systematic moral thinking can fulfil the same function simply by applying ethical concerns in the larger context beyond our immediate group-identification. All such ethical systems also have foundational assumptions that are questionable, but these do not prevent any of the approaches from stretching our moral capacities within the limits of the ideological stage.

In terms of the socio-political values identified by Haidt, this moral development is likely to correspond to a prioritization of the universal 'liberal' values of justice, care, and freedom. At the interpersonal stage, as discussed in the previous chapter, these values are present, but are not prioritized over the socially parochial ones of loyalty, authority, and purity.[4] They may well be subordinated, as for instance when people are concerned that their compatriots receive justice, but not foreigners. In contrast, by the ideological stage we are much more likely to prioritize the liberal values directly and apply them impartially (even though there will still be framing

---

3 VIII; i.7.
4 Haidt (2012).

issues as to what 'impartially' means). We may start to become more concerned with justice for oppressed people in other countries. We may defend freedom of speech even for people we disagree with. We may exercise a duty of care on the basis of humanity, even to injured enemies.

It is characteristic of the ideological stage to assume that it is the apex stage of a hierarchy, and that no development is possible beyond it – an assumption one finds constantly both in traditional religious triumphalism and in scientism. That assumption is all the more tempting when an ideology remains relatively successful in the current conditions, and of course when one is immersed in a group that continually reinforces that foundational assumption by identifying it with apparently universal rational assumptions. It can also be found in counter-ideological groups that have questioned the foundational assumptions of a given ideology, but only to *flip*[5] to their opposites, remaining within the ideological way of thinking but just with different premises. Atheism provides one obvious example of this, postmodernism another. Such counter-ideological groups can often be identified by the presence of the nirvana fallacy – that is, a traditional approach that fulfilled a given function is rejected, but no coherent alternative approach to fulfilling that function is offered in its place. Counter-ideology remains firmly within the dynamics of the ideological stage, and should not be confused with the succeeding interindividual stage.

It is at the ideological stage that metaphysics becomes visible as a phenomenon. Its visibility comes with the development of the abstract intellectual language of universality, but that increased visibility does not mean that it was not present implicitly in the earlier stages – for instance in the absolute beliefs in the rightness of the group often found at the interpersonal stage, or the absolute belief in the desirability of concrete objects found in the imperial stage. The ideological stage finally gives us the capacity to explicitly express metaphysical beliefs, and thus use them as a group-binder of a kind that can be both faster and more effective than organic social affiliation.[6] For instance, even as early as the twelfth century, the Crusaders coming together from different parts of Western

---

5   I.2.c.
6   I.5.e.

Europe had an ideological basis for working together 'against the infidel' that transcended their differing national origins.

Of course, that same group-binding creates conflict with other ideological bases, and thus starts to potentially expose us to arguments that undermine it by pointing out aspects of our experience that are inconsistent with our ideology. In *Absolutization* I detailed a whole set of defences that metaphysical belief has evolved to deal with this, including claims of absolute foundations, circular argument, infinite rationalization, the claim that metaphysics is inevitable, and the inflation of the practical importance of metaphysics in human experience.[7] All these metaphysical defence mechanisms are typical of the ideological stage of thinking, and help to prevent people advancing anywhere beyond it. They are likely to be employed to the full by those defending institutions of all kinds from individual questioning, whether these are churches, businesses, political groups, or academic networks.

Initiation of the processes that can lead us beyond the ideological stage, as with transition from the previous stages, begins with wider context. Beliefs that we previously assumed to be absolute and unquestionable become questionable – not just in a siloed space of carefully abstracted enquiry, but in a way that is allowed to influence all the other beliefs in one's life. In the transition from interpersonal to ideological stage, the unquestioned norms of the group became questionable in a wider ideological context. In the next transition, however, it is the unquestioned norms of the universal ideology that become questionable. This may begin in a predominantly intellectual way, or may consist more of gathering doubt about the emotional sustainability of the self-sufficient view of oneself one has developed whilst motivated by one's universal ideology.

The intellectual route beyond the ideological stage depends on sceptical argument. Sceptical argument, as I've previously detailed in *The Five Principles*, is argument that points out the uncertainty of all absolute claims by showing their lack of justification.[8] Uncertainty can be shown by empirical argument (pointing out the limitations of our senses, sources of information, and perspective), rational argument (pointing out the dependence of our assumptions on a network of other assumptions), or linguistic argument (pointing out

---

7   I.4.
8   II.1.a.

the lack of absolute representation in propositional language). Any of these routes, taken seriously, can suffice to show the uncertainty of any ideological foundations. The implications of sceptical argument are *not* negative, as has very often been wrongly assumed, but imply the Middle Way, and they do not undermine incrementally justified belief, only absolutizations. At some level, this point needs to be recognized for us to progress from flips within the ideological stage to the greater contextualization needed to move beyond it.

The alternative, more emotionally-oriented way that people may develop beyond the ideological stage consists in a questioning of the self-sufficient self that has been developed in dependence on it. Kegan seems to think that this route is actually more common than the sceptical route. The limit of the 'institutional balance', he writes, 'tends to show itself more clearly in the private regions of love and closeness than in the public light of work and career'.[9] As in his example of Michael mentioned above, this limit may show itself in awareness of the conflict between one's own long-term welfare and the rigidity of the persona one has adopted to fit the ideology.

A lack of deeper meaning or satisfaction, perhaps even triggering an existential 'mid-life crisis' may come to the fore, demanding a more authentic approach that pays attention to our varying embodied states and needs as well as the ideological demands we place on ourselves. Psychological approaches to the mid-life crisis have tended to emphasize our awareness of approaching death and dismay at failing to achieve life-goals that we may be attached to:[10] however, these goals are likely to be formulated in terms of our ideological framework, and our failure to achieve them is likely to show the practical inadequacy of that framework. Ideologies, being based on partial accounts of conditions, are very likely to idealize, setting goals that are not sufficiently connected to awareness of our starting points and the limitations placed on us by the conditions we work in. For example, we may have an elevated career goal that requires our own transformation to be fulfilled effectively, or we may become embittered as our political or religious goals slip further away despite our best efforts. When we fail to reach our idealizations, we are eventually forced to recognize our embodied weaknesses and limitations – a realization that may then trigger

9   Kegan (1982) p. 242.
10 Freund & Ritter (2009).

despair, flips to opposite goals, or *perhaps* a transition to a more dialectical and authentic interindividual stage.

Unfortunately, our cultural norms are currently set up to make transition beyond the ideological stage difficult and painful. We may make it partially in particular spheres – for example, responding to stress-triggered ill-health by adopting a more measured and realistic approach to work, or recognizing some of the false self-sufficiency that may be assumed in serial monogamy and compromising more fully into a more mature relationship where both partners accept each others' limitations. However, our institutions are largely set up to discourage us from following through the implications of such development by reflecting on their implications for all our underlying beliefs. Workplaces are overwhelmingly set up on an ideological basis, governed by a set of explicit or implicit values that are the basis of the organization and ensure predictable interaction within and beyond it. Individuality, creativity, authenticity, and sceptical enquiry rarely fit into that ethos, except in certain highly circumscribed situations where they are useful to the organization. The social expectations of wider public life are similarly governed by mainstream ideological consistency (and with good reason, for without that consistency democratic life is under threat). In the academic life that helps to train people for and to frame public and institutional life, scepticism is almost universally misunderstood, as I argued in *The Five Principles*.[11] That misunderstanding, that frames rigorous, even-handed questioning as negative, destructive, or ineffectual rather than a necessary feature of wider human adaptiveness, is another indicator of the entrenched cultural barriers that discourage transition beyond the ideological stage.

These barriers can help to account for the small number of adults decisively progressing beyond the ideological stage to the interindividual stage, according to the studies drawn on by Kegan. In fact the percentage of people he gives as completely reaching Stage 5 is nil – though one can take this to represent too small a proportion of the population to be picked up by the sampling represented by the studies, rather than an actual nil. However, the proportion of people showing a mixture of Stage 4 and Stage 5 responses is between 3% (in a cross-section of the wider population) and 10% (in a more educated sample).[12]

11   II.1.b-f.
12   Kegan (1994) p. 195.

This situation can lead us into some further final questions about the interindividual stage beyond the ideological. Is it a distinct stage at all? If so, what is its positive basis? Can we be reasonably justified, then, in thinking it the final stage we can achieve? My impression is that stage theorists, even including the otherwise excellent Kegan, have not really answered these questions adequately, but that the framework offered by the Middle Way can help us to clarify them. That will be the subject of the next chapter.

2.g. The Interindividual Stage

> *Summary*
>
> Each of the stages contextualizes the previous ones and addresses new conditions, extending the reinforcing and balancing loops of previous phylogenetic development. To see how the interindividual stage contextualizes the others, an embodied perspective is needed, finding our 'foundations' literally in experience which contextualizes merely abstract beliefs. The cultural entrenchment of ideological foundations can make us either deny the possibility of interindividual judgement, or alternatively make us speculate on merely abstracted further 'enlightened' stages that offer no additional context.

It may be helpful here to quickly review the overall features of stage progression that we have found throughout our tracing of individual human development. Each stage is based on a further contextualization of the assumptions that form the basis of the previous one: bodily impulses are contextualized by the imagination, the imagination by consistent concrete belief, consistent concrete belief by social trust, and social trust by consistent abstract belief. Each stage incorporates all the previous ones fully, rather than leaving their achievements behind. Transitions between stages are actuated by the recognition and integration of conflict between our habitual assumptions at our current stage, and the demands of new conditions that are not being met in the terms of those assumptions. The new stage is decisively embraced when the weight of our judgements reach a tipping point, after which the value of the new level of judgement becomes increasingly evident, and the old starts to wither away. However, our personal experience, as well as Kegan's data, suggest that those tipping points may often be unclear and messy. This fits with the asymmetricality of integration that I discussed in *The Five Principles*.[1]

This process of development at individual level also continues systemic patterns of phylogenetic development that I traced in the first section of this book. At each of the evolutionary stages traced there – the emergence of self-organizing life, sexual reproduction, multicellularity, bilaterianism, bilateral specialization, and asymmetry – the process of life asserting itself has been continued

1   II.5.g.

in reinforcing feedback loops, but new conditions have then introduced conflicts, requiring adaptive balancing loops to develop new genetic forms. The stages of individual human progression charted here are all based in the earlier fruits of this phylogenetic development, and in some respects recapitulate it in the very early growth of the zygote in the mother's womb. As we move on into the six stages of individual human development, however, each individual has to work through them anew. Now they are aided not by the strong conditioning of genes and epigenesis, but the much weaker scaffolding offered by culture and social expectation to try to ensure that genetic potential is fulfilled. As the stages go on, this socio-cultural scaffolding becomes ever weaker, so that, although the vast majority of people reach the interpersonal stage, at best around half of adults reach the ideological stage and a very small minority even partially reach the interindividual stage. At this point, then, we are at the crucial working point of the development of complexity, as it has built up from single-celled organisms to complex humanity. The question now lies in the balance, no longer a matter of history – will we make the crucial adaptations that we need to new conditions, or will we (either as a species or as individuals) fail that test? Those adaptations are not primarily ones of technology, environmental practice, or any other externality, but most basically of sufficiently complex links in individual human brains and nervous systems.

For this crucial development, the fifth or 'interindividual' stage is vital. 'Interindividual' is Kegan's term for it, based on the ways that it brings us into awareness of the individual conditions shaping our response to universal demands. The 'inter' indicates how this stage can also put us back into closer and more authentic relationship with others in the manner of the interpersonal stage, in contrast to the relative self-sufficiency, individualism, or even possibly alienation of the ideological stage. This new intensity of relationship, however, is not conducted on the group's terms, as it so often is in the interpersonal stage, but rather on the negotiated terms of an individual who faces up to all the embodied conditions making him or her a social animal, at the same time as maintaining distinct views and needs that also have universal aspirations. To do justice both to the individual situation and the universal aspirations as far as possible, a dialectical approach becomes central – which means that the Middle Way is now practised (whether or not that explicit term is used).

From the classic and dominant standpoint of the ideological stage, there can be no further stage, as scepticism is not recognized as a helpful enquiry, and embodiment is reduced to the ideologically-compatible working of metaphysics (whether mind-body dualist or monist). That standpoint is reflected in Kohlberg's scheme of moral development, the apogee of which is a Kantian stage of adopting universal moral principles determined by consistent reason. From this standpoint, there can be nothing higher than rational consistency, regardless of the fact that further argument can easily show the paradoxical cracks in this 'rational consistency'. If you believe that there is nowhere to go beyond the ideological stage, you are likely to just want to desperately keep filling in those cracks. However, this follows the pattern of excessive left-hemisphere dominance as charted by Iain McGilchrist, in which existing assumptions are constantly absolutized.

Following the general pattern of the previous stages of individual development, there is, in my judgement, a high level of justification for concluding that there is a discernible fifth or 'interindividual' stage following the ideological. This new stage contextualizes consistent abstract belief in a wider pragmatic dialectic in which our embodied experience and its limitations are effectively integrated with the use of ideological frameworks as tools of understanding for specific contexts. This stage does not reject the fruits of any of the previous ones: we still need imagination, concrete thinking, interpersonal loyalties, and consistent theory as much as ever. However, none of them offer us the whole story, and the fifth stage is thus the one in which the drawbacks of absolutization can be most clearly seen. The need for a fifth stage is pointed to particularly by the limitations of all ideological foundations, as identified by sceptical argument. Without the possibility of a fifth stage, this lack of foundation is misunderstood merely as a negative ideology – as 'nihilism'. If we pay sufficient attention to the need to recognize embodiment in the fifth stage, however, we can find a positive basis of confidence that incorporates those of the previous stages rather than rejecting them. A tipping point has probably been reached when we are no longer distressed by the supposed groundlessness of losing a secure ideology, but rather recognize the ground of our confidence as the literal embodied ground on which we stand.

We can recognize the consistent possibility of such a stage, however, to a large extent only by participating in it – at least partially and in some respects, even if we are not yet able to dwell in it consistently. Participation in it is, unfortunately, strongly discouraged by the entrenched absolutization in many aspects of culture, whichever stage prior to the fifth that absolutization is entrenched in. Such entrenched absolutizations continually reinforce the idea that we need certain fundamental abstract beliefs to function, rather than a mere bodily experience that *employs* abstract beliefs – these fundamental beliefs are metaphysical ones, most commonly defending the ideological stage, but sometimes the interpersonal, by the techniques I detailed in *Absolutization:* ad hoc or circular argument, infinite rationalization, a conviction of its own inevitability, and an inflation of its own practical significance.[2] In the face of the social pressures orchestrated by these manifestations of metaphysics, whether in the spheres of philosophy, religion, politics, education, or popular belief, we have every incentive to keep wandering confusedly on the margins of ideology rather than decisively stepping beyond it. Almost everything possible is done to *obscure* the possibility of another way of thinking beyond ideology.

In the next section of this book, I will be tracing this cultural entrenchment of absolutization in more detail through history. In its earlier manifestations, it is largely directed towards maintaining the instrumental use of absolutization at the interpersonal level, where values of loyalty, authority, and purity are likely to be absolutized. However, as we will see, more complex forms of culturally entrenched absolutization subsequently develop in more modern history to defend absolutization at the ideological level. The historical development of predominant psychological stages thus helps to determine the cultural defences that the practice of agnosticism needs to be able to resist.

Just as those locked in the ideological stage may doubt the possibility of a further stage, those prone to idealization or speculation may become attached to the possibility of further stages beyond the interindividual. Whilst, of course, one cannot rule out that possibility, at the same time there seem to be no substantial resources in human experience by which to understand it. Given that each stage contextualizes the previous ones in a further complexity, it

---

2   I.4.c–f.

is difficult to see where we would go after our abstract intellectual beliefs have been more thoroughly contextualized in an embodied perspective. Those who assume that 'higher' stages after this must necessarily be still more abstract are probably still recycling Platonic assumptions. All stages up to this point have added further contextualization, but any advances based only on the left hemisphere would reduce it. Claims about 'enlightenment' as a supreme human stage, on the other hand, tend to end up appealing, in practice, to abstracted metaphysical claims – either about the historical achievements of certain individuals, or essentialized features of human nature. Since no such supreme claim is required to justify the preceding stages,[3] we are better off sticking to the bottom-up universality[4] suggested by the stages we can relate more closely to experience.

The asymmetry of integration, too, reminds us that although there may be discernible tipping points of reframing at each stage of long-term integration up to the interindividual, those tipping points may be reached in some areas whilst remaining isolated from others, as we fail to make the synthetic links between different spheres of experience. This creates some degree of doubt even about whether we can consistently dwell at the interindividual stage, applying it to all judgements, let alone go consistently further. As noted above, reaching the interindividual stage in relation to one's personal relationships is probably more common than applying it consistently in a philosophical way. The experience of stages is almost always likely to be more ragged and approximate than our models, requiring continuing provisionality in the way we apply them.

3 See Ellis (2019) 6.e.
4 I.8.a.

3. Provisionality and Absolutization in Human Culture

Provisional and absolutized cultural features discussed in section 3

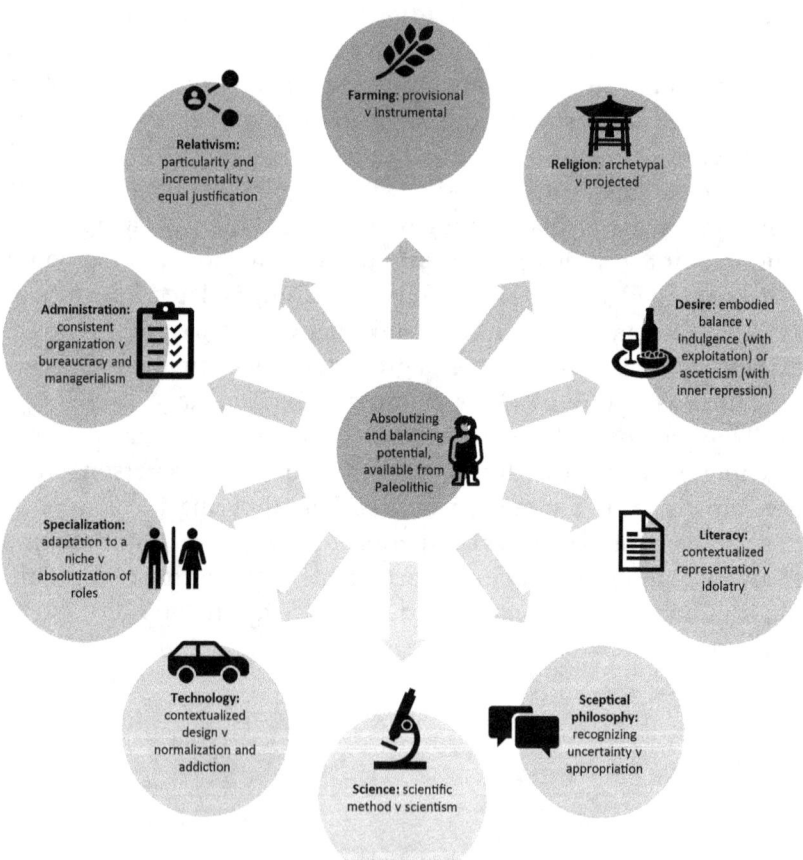

3.a. Provisionality in the Old Stone Age

> *Summary*
>
> Early humans seem to have first developed repressive abilities with representational language. However, that absolutization was not necessarily culturally entrenched in paleolithic society, because such societies varied greatly, some being egalitarian and evidently using ideological level judgement. Nor, though, should we idealize these early societies by contrast with the later effects of agriculture and the state, since the capacity for absolutization was there throughout.

To track the conditions for absolutization and the Middle Way onwards into human history, we must now turn away from the more general conditions of biology, and towards the cultural conditions that have scaffolded both human development and its failure through the ages. In human cultural history, however, we will find many of the same patterns echoed in nested conditions operating over periods of time: namely the patterns of reinforcing and balancing feedback loops, asserting and adjusting by turns. Compared to the phylogenetic patterns tracked in section 1, these patterns of human socio-cultural absolutization and integration will show a marked pattern of exponential acceleration. Compared to the similar patterns in an individual human life discussed in section 2, however, these patterns are relatively slow, although individuals of recent generations have experienced big changes in their lifetimes. In other respects, the individual development discussed in section 2 provides the background conditioning that can help us to understand cultural changes. As we will see, collective judgements can reflect the psychological stages that shape individual judgements, even if the group as a whole cannot be said to be at a determinate stage.

Sections 3 and 4 of this book will only attempt to track human history in an extremely loose fashion. Although I begin with the issues of early human history, we will soon find it easier to consider the factors that have produced human absolutization in terms of related strands of human culture and experience, each tracked separately through historical time. The factors that I will consider in section 3 all involve the hijacking of adaptive human features, whereby a cultural change that started out as a balancing feedback loop turns into a reinforcing one. In some respects, then, it is a history of absolutization, though in others it is also a history of positive adaptations that preceded those absolutizations. The negative emphasis in this section will be balanced out by a positive emphasis in section 4, where I will be tracking the development of practices that we can use in the service of integration and the Middle Way, potentially thus turning the reinforcing feedback loops back into balancing ones.

I begin, then, with questions about the earliest conditions in which humans emerged. As I discussed in 1.i, the capacity of each hemisphere to suppress or repress the other for practical purposes seems to be crucial in the development of distinctive human abilities. The left hemisphere represses (and thus absolutizes) without sufficient contextual awareness, but the right merely suppresses, maintaining provisionality. Group reinforcement of a kind found in many animals can strengthen that repression or suppression through our constant awareness of whether others share the same states. Left-hemisphere repression of the right has a rigidity lacked by the provisional right-hemisphere suppression of the left, because it leaves us dominated by merely conceptual belief and resistant to any alteration in that belief. The development of left-hemisphere propositional representation in early human history seems to have made a crucial difference in our vulnerability to this repressive rigidity.

This leaves us with some difficult questions to try to resolve with reference to early human origins. Firstly, when and why did we develop this repressive ability? This is a complex problem in early human anthropology on which I can only suggest the outline of a helpful approach. Secondly, how and when did this repressive ability start to get culturally entrenched enough to create serious adaptive difficulties for us? This is much more of a historical question, on which I think it is possible to make more progress, at least by ruling out certain dogmatic assumptions.

The first question – when and why did we develop repressive abilities? – seems to be closely tied to debates about the origins of human language. Our language skills as a species combine capabilities in the brain, in the vocal tract, and in social interaction. These can be helpfully divided between language capabilities in a broad sense, that show an incremental development from earlier animals, and language capabilities in a narrow sense – that of distinctively human representation.[1] Scholars remain divided on the timeline for these key elements coming together to allow distinctively human communication skills. This may have taken place during the development of *Homo erectus* (between 2 million and 500,000 years ago), or as recently as 100,000 years ago in the development of *Homo sapiens* (when Philip Liebermann places key physical changes in the vocal tract[2]). It is in this 'narrow sense' that we can probably trace both repressive and suppressive skills in the human brain, because it includes *recursion*, the ability to combine propositional elements in relation to each other. Hauser, Chomsky, and Fitch note that this ability 'generates internal representations and maps them into the sensory-motor interface by the phonological system, and into the conceptual-intentional interface by the formal semantic system.'[3] When we repress, it seems likely that we hold some of these representations in awareness whilst attempting to eliminate alternatives, but when we suppress, we focus on certain representations for practical purposes whilst retaining some base awareness of the alternatives. As I have been arguing throughout this series, representationalism as the still-dominant view of meaning makes the terrible mistake of identifying representation with meaning. The dawn of propositional representation in history is clearly *not* the dawn of human meaning. However, it *is* the dawn of human absolutization, alongside the equally vital dawn of human provisionality. We remain very unclear about even approximately when this dawn occurred, but we can be reasonably clear about the conditions it required.

It is the second, cultural and historical question, though, that tells us how that capacity for absolutization became effectively activated, as it was passed on and employed socially. It is a question much more often answered in socio-political than in psychological or neuroscientific terms, and which forms the basis of a vexed debate about how inequality became entrenched in human societies, and whether or not it is avoidable. I see this debate about inequality as merely one strand in the wider question of how power started to be exercised rigidly within the individual human psyche as well as between humans and their groups, but nevertheless it can

---

1 Hauser, Chomsky, & Fitch (2002).
2 Liebermann (2006).
3 Hauser, Chomsky, & Fitch (2002) p. 1571.

make a helpful contribution to our understanding of the history of absolutization.

This debate has recently been profoundly reframed by the publication of *The Dawn of Everything* by David Graeber and David Wengrow,[4] a book whose insights and evidence I will be drawing on extensively in this and the following chapter. This book sets out with an implicitly Middle Way purpose: to navigate between the two dominant dogmas that shape people's assumptions about pre-agricultural societies. These dogmas are typified by Rousseau on the one hand – the idealization of the 'noble savage', and by Hobbes on the other – the demonization of the early human lifestyle as 'nasty, brutish and short'. Graeber and Wengrow argue that the idealization of an egalitarian, ecologically-adapted 'primitive communism' originates in Native American critiques of European civilization,[5] but point out the interdependent oversimplification of a complex picture that is shared between this attitude and the opposing Hobbesian one. When Lockean arguments about property are added to the Hobbesian dismissal, claiming that the basis of property rights is the *usage* of natural resources, it becomes clearer that pre-agricultural societies have often been colonially dispossessed of their lands with the rationalization that they are not 'using' them in ways recognized by Europeans.[6] An entrenched over-simplified view of pre-agricultural societies thus binds together philosophical assumptions with historically narrow socio-political motives. To get beyond the simplistic dichotomies this has left us with, we need rigorous even-handed agnosticism about both the idealized and the Hobbesian beliefs about early human societies, together with an ever-provisional examination of the evidence.

Graeber and Wengrow point out that pre-agricultural societies were immensely varied, not only due to a wide variety of environments, but also to very varied cultural responses to those environments. Whilst some pre-agricultural societies, such as the Calusa of western Florida, had very rigid and hierarchized social structures of the kind we're more likely to associate with neolithic agricultural civilizations,[7] a good number also had a much more flexible and provisional 'egalitarian' structure. This variation suggests that

4   Graeber & Wengrow (2021).
5   Ibid. ch. 2.
6   Ibid. pp. 148ff.
7   Ibid. pp. 150ff.

both absolutization and provisionality have been aspects of human experience since well back into the palaeolithic era. Dogma, power, and repression (as well as their antidotes) were thus not merely products of agriculture and its effects on human society, but more general features of human psychology that merely get culturally entrenched in particular circumstances.

The 'egalitarian' structure that made paleolithic societies often more provisional was one of 'substantive' rather than formal freedoms.[8] As I interpret this (using the socio-political values from Haidt's research discussed above in 2.e), this implies that personal freedom was possible to a large extent because of the effective balancing of that freedom in a complex social system with other values. For instance, many Native American and Australian Aboriginal hunter-gatherers had the freedom to travel widely, accommodated by values of care for strangers shared by many other tribes. This 'substantive' freedom actually allowed considerable cosmopolitanism in many hunter-gatherer societies, of a kind that was only later restricted.[9] If each individual or group had merely asserted their formal 'freedom' to restrict entry to their lands, this would have conflicted with the freedom to travel. Similarly with the value of equality or justice: the Wendat group of Native Americans in modern Canada, for instance, had chiefs, but nobody was obliged to obey them, their social status and authority being well contextualized by other values.[10] Plains Indians have been observed to carefully rotate their leaders and limit their authority by using a ritual context.[11] Jared Diamond gives a modern example from New Guinea of ostensible followers actively thwarting the will of a tribal leader, without what we might later call political rebellion. As he writes, 'Some tribes have a "big man" who functions as a weak leader, but he leads only by his powers of persuasion and personality rather than by recognized authority.'[12]

The Wendat, too, seem to have had advanced powers of discussion for the resolution of political issues, as well as widespread tolerance of very different views and eccentric lifestyles. This is strongly exemplified by the figure of Kandiaronk, a Wendat thinker who

8   Ibid. p. 131.
9   Ibid. pp. 121ff.
10   Ibid. pp. 52ff.
11   Ibid. p. 110.
12   Diamond (2012) p. 15.

launched an influential critique of the assumptions of eighteenth-century French society, mediated by the Frenchman Lahontan.[13] These skills seem to display a level of judgement equivalent to at least the ideological stage in Kegan's theory as discussed in the previous section, with universal and highly contextualized values at odds with the more typical view we might have of hunter-gatherers as restricted to the narrow assumptions of their tribe. Hunter-gatherers might at first sight seem to be the paradigm case of a *traditional* society, but, as we will see later, it is probably the conditions of early agricultural society that did most to help produce the association of traditional societies with interpersonal rather than ideological stage judgement and with more conservative, group-orientated, values. In some respects, pre-agricultural societies have often been *pre*-traditional.

The definition of a paleolithic or foraging society can also introduce false dichotomies into a complex picture. Some largely foraging societies have also had gardens for limited agriculture. Those by the sea, lakes, and rivers may have had fishing as a major source of food – a type of 'foraging' still practised in modern society. Some 'foragers' have extended their 'foraging' into raiding the food, people, or other resources of other societies, and if these other societies were agricultural they might then thus start living on the proceeds of agriculture at second hand without doing it themselves[14] (see 3.d below for more about the implications of exploitation practices). Historians thus talk of the 'mesolithic' as the extensive grey zone between the paleolithic and the neolithic phases of development.

Some groups in this liminal position have also moved seasonally between foraging and agricultural lives. Claude Lévi-Strauss documented one of these in the Mato Grosso of early twentieth century Brazil – the Nambikwara – who practised horticulture in settled villages during the rainy season, but foraging in dispersed bands the rest of the year.[15] Here not only the way of life of these people changed seasonally, but also their entire political system, with chiefs operating during the settled period who had no authority during the dispersed period. Clearly, such customs provide an ever-renewed contextual awareness and provisionality for any claimed authority, and must contribute to the cultural influence on

13 Graeber & Wengrow (2021) pp. 48–59.
14 Ibid. pp. 188–9.
15 Ibid. pp. 98–101.

many judgements. Such mobility can give a variation of experience and thus of potential beliefs, drawn from a wider integration of meaning than we would expect with entirely settled people.

Against this, we should set the evidence that paleolithic societies were as prone to absolutization as others. One way of judging this is from rates of violent death, given the high likelihood that violence in hunter-gatherer societies was created by unnecessary disputes over, say, land or mates, that were fuelled by proliferating mental states and might otherwise have been resolved peacefully. Violence is disinhibited socio-political conflict, that cannot be initiated without conflict being present in the aggressor,[16] even if that inner conflict has been exacerbated by external stresses (such as competition for scarce food). Steven Pinker presents evidence, based on estimates from bodies that show evidence of violent death relative to overall estimated population levels, that rates of violent death in paleolithic and mesolithic societies are far higher than those in later state societies.[17] There is, of course, the danger here that the bodies discovered were in some way selected for burial because of the same factors that produced the violent death, but Pinker nevertheless uses a wide range of figures from a wide range of sources and should be required reading for those who over-Romanticize the past. He reminds us, if nothing else, of the complexity of the picture. Along with the civilized discourse in some hunter-gatherer societies, there has also been frequent inter-tribal warfare, along with torture and violent cannibalism at times.[18]

All this shows, not that pre-agricultural societies were as a whole either superior or inferior to later ones, but that we cannot assume more restrictive levels of judgement without further evidence on the conditions at work in particular circumstances. As we will see on advancing into the coming of agriculture, there are indeed many conditions that accompanied it that have helped to create much more rigid socio-cultural conditions, thus also at least incentivizing absolutized thinking and limitations in stage development. These reinforcing feedback loops in early human society, however, were later balanced out by further waves of provisionality. The ways that neolithic conditions provided the conditions for absolutization, however, should not lead us to idealize paleolithic conditions

16 I.5.a (pp. 135–6).
17 Pinker (2011) pp. 57–67.
18 Ibid. pp. 51–6.

in general by contrast. Some inhabitants of the Old Stone Age do indeed seem to have been noble savages. The lives of others seem to have been nasty, brutish, and short. However, the vast majority have probably been somewhere in between.

3.b. The Two Faces of Farming

> *Summary*
>
> Contrary to the widespread narrative, there is now good evidence that farming did not begin in the intensive, highly instrumental forms that have created such increasingly damaging reinforcing feedback loops since the neolithic. Instead, there was probably a 3000-year period of provisional farming that was liminal, flexible, diverse, egalitarian, and relatively integrable with the ecosystem. Even more intensive river-valley farming and cities did not immediately produce absolutized social structures: these instead spread from marginal areas where exploitative groups had taken over.

The most widely recognized face of farming is the one that created the 'neolithic' of the account that has been unchallenged in its dominance until recently: that is, the large-scale and systematic appropriation of land by humans to grow food or other useful products for themselves, or to rear animals for the same purposes. That appropriation, according to that account and its implications, involved both a shift towards greater instrumentality in our approach to the world (as we moulded significant portions of our environment for our own purposes), and the increasing production of surpluses that could support a population of specialists (craftspeople, traders, priests) and/or parasitic land-holders. This kind of farming set the stage for the development of urban civilization, and was followed by an accelerating feedback loop of ever-intensifying exploitation, not only of our wider environment, but also of animals and often of other humans. With the crises of climate change, as well as of biodiversity and of many types of resource use, we are now seemingly approaching the apex at which that reinforcing feedback loop becomes most frenetic and also most fragile – a point that can only precede the collapse of the system created by agriculture.

In the face of this overwhelmingly instrumental, left-hemisphere dominant, face of farming, it is easy to ignore the possibility that farming could be otherwise, despite the campaigning of various marginal groups to bring our attention to other possibilities. Some of these alternatives are marginal adaptations to slow down the juggernaut, such as organic farming or no-till, but others offer major alternative visions of how we can produce food, such as veganism, permaculture, agroforestry, and re-wilding. The new, well-evidenced interpretations of Graeber and Wengrow of the origins of farming, however, make it clear that a more provisional, experimental, and sustainable form of farming is not only part of its history, but was also the dominant face of farming for many thousands of years. There was no 'agricultural revolution', they point out: rather there was a very long period of tinkering and flexible adaptation, during which agriculture was just one of many possible strategies for getting food. The process of 'agricultural revolution' was rather an age in itself – of around 3000 years between around 10,000 and 7000 BCE.[1]

Lest this should be confused with 'back-to-nature' Romanticism, I want to point out here that such flexible and provisional farming, whether in past or present, has nothing to do with 'living in harmony with nature' in the absolute and idealized sense in which such language is often used. Though more sustainable methods of agriculture clearly involve *less* modification to our environment, and thus far less danger of fragility and system destruction than more highly instrumental forms of agriculture, both involve modification of the environment for human benefit.

The key difference between the two faces of farming should not be deduced from a metaphysical dualism between 'natural' and 'unnatural' approaches, but rather understood in terms of the differences of human judgement applied: are we treating our environment with sufficient provisionality to allow wider awareness of its systems and processes to be taken into account – or are we (literally) 'ploughing on regardless', justified by absolutized abstraction? There can be no such thing as 'natural farming', but there can clearly be 'provisional farming', of a kind evidently practised by our remote ancestors. How far we can adopt such provisional farming today, whilst feeding populations that have been unsustainably boosted by instrumental farming, is a massively important ethical and political issue – perhaps the most important issue of our time, but one I will return to in later volumes of this series. For the moment I need to focus only on the historical picture.

The evidence that the development of farming was initially so provisional can be seen in its location, liminality, flexibility, interdependence, diversity, and egalitarian uses of land. To start with, the dominant theory of 'agricultural revolution' assumes that farming

---

1   Graeber & Wengrow (2021) p. 234.

started in the most resource-rich areas, such as river valleys, where large-scale cooperative activity could produce big surpluses of food most easily. However, the evidence points much more to the very gradual development of farming in relatively resource-poor areas, where it could help to supplement the diverse diet of people still largely dependent on foraging. In the Middle East, this points to the upland areas rather than to the river valleys.[2]

It was able to develop in those areas precisely *because* it was liminal, on the edge of hunter-gatherer needs rather than requiring a great deal of risky time and effort. One early form of crop-farming, for instance, was flood-retreat farming, where crops were sown on ground left already clear and fertile by a retreating lake or river.[3] Rather than taming animals completely, they could be corralled in a way that made them easier to hunt. Where land was more intensively managed, it was on a very small scale – gardens rather than fields.[4] Agriculture remained interdependent with hunting and gathering in all the areas where it developed, but may also have often developed in co-dependence with trade, so that some small degree of specialization in producing food allowed a small surplus that could be exchanged for otherwise unobtainable goods from other zones.[5]

The flexibility of early farming can be illustrated by the ways that it changed to adapt to new environments. As Polynesians gradually colonized new islands across the Pacific, for instance, they carried with them an agricultural tradition that had its roots in China and had spread via Taiwan and the Philippines. However, they rapidly abandoned the rice and millet cultivation that was the staple in China, but unsuited to the tropical climate of Polynesia, at the same time keeping the pigs and chickens. In the process they also developed new, better-adapted crops such as taro.[6] More generally, the model of agriculture developing only from a few key points (such as Mesopotamia, China, and Mesoamerica) from which it spread elsewhere has also been challenged by more recent evidence of its independent development in at least twenty different areas of the world. For instance, in the Eastern Woodlands of North America,

2   Ibid. pp. 225–9.
3   Ibid. pp. 235–6.
4   Ibid. p. 260.
5   Ibid. pp. 225–9.
6   Ibid. pp. 265–6.

Native Americans raised goosefoot, sunflower, and sumpweed as crops, long before the introduction of maize from Mesoamerica, let alone Old World crops such as wheat.[7]

Nor did agriculture necessarily require the loss of the relatively egalitarian structures often used by hunter-gatherers, and discussed in the previous chapter, even when it was pursued more intensively. As Graeber and Wengrow write, 'Communal tenure, "open-field" principles, periodic redistribution of plots and co-operative management of pasture are not particularly exceptional, and were often practised for centuries in the same locations.'[8] It is only when the demands of instrumental efficiency begin to override those of social solidarity and flexibility that consolidated land-holding begins to be applied. Such instrumental efficiency is only required when there are greater populations to feed or more specialized and/or exploitative elements in society to maintain, but it is purchased at the cost of sustainability (both ecological and social) and biodiversity.

The forest garden, as an alternative approach, may be very ancient, and particularly seems to have been employed by those who perhaps actively chose not to go in for full field agriculture: Graeber and Wengrow discuss the example of forest gardens in the Amazon region.[9] The forest garden integrates perennial food plants of differing sizes into a single sustainable system that is nevertheless geared towards food production for humans, and requires little labour to maintain once it has been set up. A 2003 study into traditional forest gardens in Sulawesi found that they provided between 57% and 77% of their owner's necessary income, and were particularly valuable as a standby in hard times when other sources of income or sustenance failed.[10] It is easy to imagine the earliest farmers in many parts of the world (especially the tropics) adopting a similar system, probably modifying a small area of existing forest rather than planting a new forest garden from scratch: the results being liminal, flexible, interdependent, and biodiverse. They may not be a complete solution to human food needs in the future, but are surely an important element in the re-provisionalizing of human agriculture in imitation of those early times.

---

7   Ibid. pp. 252–4.
8   Ibid. p. 250.
9   Ibid. p. 270.
10   Brodbeck et al. (2003).

In their relationships with animals, the full agricultural model requires humans to become 'both the primary carers for and consumers of other species,'[11] but more provisional models may only involve one of these, fully or partially. We can 'keep' animals, or help to sustain them to varying extents, without eating them, whether this provides us directly with useful products (like sheep wool), merely delights us, or contributes to the local ecological system (as in the housing of bees). On the other hand, we can eat wild animals without keeping them, provided we can maintain this at sustainable levels. Both keeping animals *and* eating them, on the other hand, requires a marked psychological shift towards instrumentality, in which our recognition of a familiar animal as a living companion has to be alienated and repressed to allow slaughter. In hemispheric terms, our right-hemisphere recognition of other living creatures has to be forcibly replaced with a left-hemisphere dominant view of them as an object. The cultural maintenance of our ability to engage in such instrumental repression is probably the starting point for the absolutized ideology of *carnism*, which Melanie Joy has identified as requiring the assumption that meat consumption is natural and necessary, accompanied by 'psychic numbing'.[12]

In discussing the provisional agriculture in the indigenous people of the Amazon region, Graeber and Wengrow point out that

> We are dealing here with people who possess all the requisite ecological skills to raise crops and livestock, but who nevertheless pull back from the threshold, maintaining a careful balancing act between forager and farmer.[13]

There could be a variety of reasons for 'pulling back from the threshold' like this. Graeber and Wengrow show how some ancient societies may have done so mainly to differentiate themselves from their neighbours. The need to develop carnist ideology may be a factor for those used only to either keeping animals or hunting individually unfamiliar ones. In the context of a hunter-gatherer society, there may be an active appreciation of the ways that a further adoption of systematic agriculture reduces optionality and adaptability.[14]

The drawbacks of that further adoption can also be illustrated historically in the examples of several early neolithic villages that

---

11 Graeber & Wengrow (2021) p. 268.
12 Joy (2003).
13 Graeber & Wengrow (2021) p. 268.
14 II.2.a.

have been excavated in Austria and Germany, dating from between 5500 and 4500 BCE. These people seem to have migrated inland into Central Europe from the coastlands, where there had been a much wider variety of food available from foraging. Entering the Central European plains, these people seem to have become over-dependent on grain cultivation, with few other sources of food available to them, which had the effect of weakening them in the longer term. They were then attacked and overwhelmed by rival foraging communities.[15] Graeber and Wengrow rightly call this 'a neolithic cautionary tale': far from being an instant ticket to prosperity, absolutized farming could become an uncompetitive cul-de-sac in changing conditions. In addition, dependency on grain agriculture could lead to a decline in the diversity of nutritional sources and a constant demand for back-breaking labour, having a large impact on the health and well-being of people. The adoption of corn as the basis for Native American diets in the Illinois and Ohio River Valleys seems to have led to markedly increased adult and infant mortality, widespread epidemic and degenerative diseases, rampant tooth decay, and anaemia.[16]

Nevertheless, of course, later societies based on more intensive agriculture did succeed. Between 5000 and 4000 BCE we find the first evidence of cities developing in Mesopotamia. Historians[17] have long noted the reinforcing feedback loop of interacting factors that made them possible: an exponentially developing availability of food, interacting with an expanding sedentary population, linked in turn with developments in technology, trade, and social organization. It seems that the sheer quantity of food that could be produced by agriculture, and the resulting expansion in quantity of human societies, outweighed any losses in quality of life. Of course, in the longer term, agricultural societies did help to create some of the conditions that would allow further balancing developments to fix some of these drawbacks. For instance, the surpluses much later allowed a development of specialized medicine that could help to improve health. However, this was only at the wider cost of accelerating exploitation of the environment as a whole.

Even as agriculture began to catch people in loops of labour, diet, population growth, and environmental exploitation, however, the

15   Graeber & Wengrow (2021) pp. 260-3.
16   Diamond (1991) pp. 168-9.
17   E.g. Christian (2005) ch. 8.

further absolutization that is often associated with the rise of agriculture – of hierarchized authoritarian social structures – did not yet necessarily arrive. Graeber and Wengrow collect fascinating evidence of relatively egalitarian republican socio-political structures in a range of early cities. In early Uruk and other Mesopotamian cities in the fourth and early third millennium BCE, there were massive public works, such as irrigation dykes constructed by the labour of corvées (work gangs) – however, people of all classes contributed to these in a spirit of festivity.[18] There are no signs of Royal Palaces in the archaeological evidence of that time, but citizens' assemblies were continuing to operate even after kings did take over in Sumeria.[19] A similar picture can be found in a range of other neolithic cities, such as in the egalitarian structure of those excavated in the Ukraine,[20] the presence of a Great Bath rather than a palace as the communal focus in the Indus Valley city of Mohenjodaro,[21] and even the presence of republican government in the cities of Teotihuacan and Tlaxcala in pre-Columbian Mexico.[22]

This is a widespread but not a universal pattern. Where paleolithic forms of egalitarian social organization existed, they seem not to have immediately given way to centralized authority in the first civilizations just because more people were clustered together. Rather, people were often able to organize themselves by continuing to distribute authority and reach decisions by consensus. Such a communal decision-making structure, whether one calls it 'democracy' or not (see 4.1 below), continues to have the longer-term strengths of democracy, because of the degree of socio-political integration it allows. Where divergent opinions are taken into account, a wider range of conditions is addressed, and more integrated motives are maintained for all the individuals involved. In the longer term, then, adaptability is maintained, and balancing feedback loops become part of the communal operation of the group. However, such communities are vulnerable to the violent incursion of groups bound together by absolutization, giving them a quick fix for dominance in certain restricted conditions.

---

18   Graeber & Wengrow (2021) pp. 298–301.
19   Ibid. pp. 301–3.
20   Ibid. pp. 288–97.
21   Ibid. pp. 313–21.
22   Ibid. pp. 328–58.

In the case of Sumeria, that absolutizing wave seems to have come from the north, and the first evidence we find of it is in the city of Arslanstepe in eastern Turkey. Here the communal temple was replaced by an exclusive palace, accompanied by bronze age weapons, and signalling the spread of a warrior aristocracy from around 3100 BCE.[23] It seems likely that at that stage, the ideological level of judgement that may have arisen in the psychological development of some of those involved in republican self-government was replaced by a widespread reversion to group-based interpersonal thinking. Such thinking only has to activate the coordinated left-hemisphere dominance of the group, evolutionarily shared with other animals and discussed in 1.h, to sideline the more subtle contextualizations of cooperative decision-making in favour of the immediate interests of the ruling group. The history of such exploitative societies is discussed more fully below in 3.d.

Perhaps some will feel that I have drawn excessively on Graeber and Wengrow's work in this chapter and the previous one. However, it offers a critical perspective on previous historical assumptions that is as yet little developed elsewhere. Like any other empirical claims based on evidence, it should be taken provisionally, and may be refined or superseded by further evidence. However, the importance of their work for our understanding of the Middle Way in early human history should not be underestimated. They offer historical evidence that there is nothing inevitable about the dominance of absolutization in human social experience. They show that the ability to make judgements with a fair degree of provisionality, taking into account a range of conditions, is nothing new: as I shall discuss further in 3.g below, scientific method is not such an exclusively recent development. They also show, as I discussed in 1.i above, that humans not only *can* relate on the basis of solidarity rather than group-repression, but actually have been doing so, at least to some extent, for thousands of years.

In the remaining chapters of this section, we will find a similarity of structure to the one observed in this chapter with regard to farming: namely, the initial presence of a provisional human structure that addresses a wide range of conditions and needs, but is then hijacked by absolutization so that it tends to focus only on much shorter-term conditions. As in the case of farming, where the

---

23 Ibid. pp. 310-13.

entrenched view has seen absolutization as inevitable, this turns out not to be the case at all, but merely a matter of our framing of the matter. We will find this in the cases of religion, of wealth and economic relationships, of the use of written language, of sceptical philosophy, of scientific method, of the use of technology, of specialization, and of bureaucratization. In each case, the history will only be chronological within the treatment of a given topic, and of course (given constraints of space), the historical details will need to be very selective.

3.c. Religious Archetypes and their Projection

> *Summary*
>
> Early human religion is often interpreted in terms of 'beliefs', fuelling reductionist or dogmatic interpretations of it, but this is not obvious from the evidence. Early religion is archetypal meaning, with probably a mixture of integrated and projected use of those archetypes. At some point in the paleolithic, many groups became more insular, leading to the interpersonal level judgements of ethnic religion. In the Axial Age, this expanded into ideological universality, giving new awareness, but also metaphysical expressions to projected religion. The interindividual judgement offered in the Buddha's Middle Way is needed to go fully beyond this.

Religion is perhaps the most deeply rooted and the most fraught of the areas of human activity that began as a practical adaptation, but have been hijacked. That is particularly because the hijacking of religion as dogma has become so culturally entrenched that it is commonly understood as completely definitional. 'Religion', not only popularly but even for most scholars, is almost equivalent to 'religious beliefs'. The Oxford English dictionary begins its definition of religion with 'the belief in and worship of a superhuman controlling power, especially a personal God or gods'. Even scholars with more nuanced definitions that take into account the embodied context of religious practices still tend to place 'belief' at the core: for instance, Gregory Wightman, in a recent book on religion in the paleolithic, writes 'I regard a "religion" as a shared body of beliefs, understandings, and experiences in respect of entities that possess and exercise power.'[1] Here, even the understandings and experiences of religion are still interpreted as understandings and experiences *of* projected supernatural entities, assuming a belief model of religion before we even start to examine the topic. However, defining religion in terms of its projections is a bit like defining government as necessarily oppressive or wealth as necessarily exploitative – it places an unquestioned ideological framing around our very understanding of the phenomenon, so as to observe it from a distance as what those funny dogmatic people do, whilst neglecting the religious dimension of all human experience.

1   Wightman (2015) p. 11.

My account of religion as functional inspiration through archetypes is already given in detail in my earlier book *Archetypes in Religion and Beyond*,[2] so I am not going to make any attempt to do more than briefly summarize it here. There I suggested that religion is the practice of integrating the archetypes – where archetypes take the form of varied symbols but share the function of integrating our awareness over time, and thus maintaining sustainable goals, relationships, threat-responses, and engagement with potential. When religion succeeds in performing that function, through providing symbols that inspire us over time, we can call it *practical religion*, but all too often that function has been hijacked by projection, by which we cease to take responsibility for our practical integration, and instead seek absolutizing shortcuts. Those shortcuts, in religion, often consist in 'supernatural' projections[3] – that is, gods acting in the world beyond ourselves. Religion dominated by such shortcuts (producing 'religious belief') I call *dogmatic religion*. It is very important to distinguish dogmatic religion from religion in general for what are at root practical and moral reasons: so that we can engage the resources of religious tradition to help us practise the adaptive Middle Way, creating balancing feedback loops in our judgements rather than further reinforcing ones.

What I am concerned with here, instead, is the implications of this approach for the history of religion. If there is one wider message one can get from Graeber and Wengrow's work, it is that views of early human history are hugely dependent on prior assumptions. To a large extent, when sorting through ambiguous archaeological or historical evidence, we find what we are looking for and interpret it in terms of our existing frames of assumption: thus, for instance, that the Mesoamerican city of Tlaxcala was a republic was a point that was hiding in plain sight, but has evidently been ignored by historians who assumed that Mesoamerican governance must by definition have been authoritarian, and thus any mention of this point was a projection of European classical models onto Pre-Columbian Mesoamerica.[4] Similarly, when historians of early religion tell us that such-and-such a group 'believed' this or that, the lack of necessity for this assumption might well be hiding in plain

2 Ellis (2022).
3 Note that 'supernatural' here is *not* intended to contrast with 'natural', only with 'provisional'.
4 Graeber & Wengrow (2021) pp. 346–58.

sight. No doubt it was sometimes correct, but to stop making the assumption that 'belief' in deities that 'exercise power' is constitutive of early religion, one only needs to offer a plausible account of why we should stop making that assumption, and why the evidence does not dictate it in every case. In that way we can add our view of religion to a wider understanding that human beings are indeed capable of provisionality in a wide range of situations, and that there is nothing inevitable about metaphysical belief.

As representative of common attitudes to early religion in the social sciences, let's take a 2006 article by Matt Rossano.[5] Reviewing a range of earlier studies, Rossano identifies the earliest sources of religion in the early to mid paleolithic, around 300,000 to 150,000 years ago, where it takes the form, he says, of the inducing of ecstatic states, possibly accompanied by rituals in deep caves, for the purpose of social bonding. He does not regard this as yet fully religious, because it gives no indication of supernatural beliefs. After about 150,000 years ago, however, shamanistic healing rituals appear. He sees these as the vector by which supernatural causal beliefs developed as an extension of the causal beliefs formed by early humans in relation to their everyday environments and others. Foremost amongst such supernatural causal beliefs, he writes, is the belief that ancestors can continue after death, influencing the lives of the living, which he sees as typical of upper paleolithic religion from about 35,000 years ago. Grave goods are thus taken to provide clear evidence of supernatural beliefs.

Here we have a whole range of questionable assumptions, not about the evidence of early humans, but about its interpretation. It is assumed that ecstatic states, and religious experience in general, are solely for the purposes of social bonding – which is applying dogmatic social reductionism to what is a *meaningful experience* for each individual. Experiences did not cease to be meaningful just because they happened to paleolithic people. Such experiences may well have the effect of opening our contextual awareness to make it easier to overcome conflict and find solidarity with others, but that does not mean that religious experience is reducible to such a social effect. It is assumed that to be 'religious', paleolithic people have to believe in supernatural entities – yet the evidence cited gives no indication of this unless we read back from the post-Enlightenment

---

5 Rossano (2006).

obsession with 'belief' as the basis of religion. Grave goods, for instance, may tell us something about the *meanings* associated with a dead person, but the motives for placing them in a grave may be multifarious: because of a strong association with the dead person, a strong association with death, a wish to keep the goods apart from living use, or a wish to follow the wishes of the dead person who was attached to these objects. It does not necessarily tell us anything about causal processes that are 'believed' to follow from the placing of grave goods in a grave, whether to the benefit of the living, or to the benefit of the dead person who is 'believed' to continue in some way. We simply have no justification for assuming such things, which are the product of a modern post-Christian account of religion as 'belief'. Like the other common assumptions about paleolithic and neolithic peoples discussed in the last two chapters, they may be correct in some cases, but are most unlikely to be correct in all.

If, instead of applying either social reductionism or the dogmas of any one given religious tradition, we consider the functions of the earliest signs of religion as *meaning*, it is possible to develop at least a plausible account of it that may well have been correct in many cases. Paleolithic peoples often made use of hallucinogenic plants to induce ecstatic states for probably very similar reasons to those who followed them in the 1960s: because those states themselves opened their awareness to something larger, which was itself spiritually fulfilling. Symbolism then developed – for example in cave art, in totemism, or in ritual – that may then have become *associated* with those enlarging experiences. Rossano mentions the use of red ochre for symbolic purposes – a substance that has no other obvious purpose, but that seems to have been used up to 1 million years ago.[6] He also writes 'Many items recovered from late lower and middle paleolithic sites (from 300,000 to 35,000 years ago) contain what appear to be intentional zigzagging, parallel, or radiating markings that may represent the entoptic experiences[7] of altered states of consciousness.'[8] Such symbols triggered a larger perspective that linked those people in awareness with themselves over time and with others, having an integrative effect.

---

6 Ibid. p. 352.
7 That is, experience of visual phenomena that are produced within the eye.
8 Ibid.

That larger perspective may have helped those same people in practical judgements of all kinds, including causal ones. For instance, cave art can include figures of animals and indeed hunters, alongside other mysterious symbols. In the case of a particular instance of cave art from Cueva Manos, Argentina, hunters and animals are accompanied by outlines of human hands surrounded by colour.[9] Is this evidence that the hunters believed that their gods, or ancestors, would help them in the hunt (perhaps persuaded by the right rituals or grave goods)? We can't rule that out, of course, but we could also just as easily conclude that the hunters were more likely to make better judgements in their hunt if they stayed in touch with the larger perspective offered by their spiritual experience – as perhaps symbolized by the hands surrounded by colour. The choice of interpretation involves applying either our own left-hemisphere instrumental priorities (that the hunters wanted to 'use' the supernatural directly to help them achieve their objectives) or a wider right-hemisphere perspective (that religious experience was just part of the wider context in which the hunt, like everything else, took place).

Of course, it is also likely that the projection of religious archetypes into supernatural entities goes back to the paleolithic period. As we have seen in the last two chapters, the distinctions between paleolithic and neolithic ways of living have often been over-stated, indicating that if we find a phenomenon clearly present in early agricultural societies, it is likely to taper back well before those societies. Such phenomena as shamanic healing and ancestor worship in the paleolithic *may also* have offered tempting shortcuts to wish-fulfilment, shaped by biases, despite those shortcuts not necessarily being the main value of them. In such a shortcut, absolutization kicks in because we believe the products of bias.[10] Again, though, such belief is not inevitable.

The distinction between the earliest forms of absolutization in human history, and the later ones associated with the development of the world's main religious traditions, is much more likely to be one between interpersonal and ideological ways of thinking than one between absolutization and provisionality. As we have seen (in 3.a), ideological ways of thinking with an aspiration to universality do

---

9   Previously reproduced, and discussed in a different context, in Ellis (2022) p. 148.

10   I.5.c.

seem to have arisen in paleolithic or mesolithic societies. However, the cosmopolitan phase of paleolithic life seems to have been succeeded by a more restrictive and parochial phase. It is then that 'traditional society', whether hunter-gathering or agricultural, appears, with its emphasis on the interpersonal and the values of the tribe as the foremost values. Jared Diamond gives expression to the limitations of this way of life:

> Rarely or never do members of small-scale societies encounter strangers, because it's suicidal to travel into an unfamiliar area to whose inhabitants you are unknown and completely unrelated. If you do happen to encounter a stranger in your territory, you have to presume that the person is dangerous.[11]

Given this culturally entrenched bias against strangeness and difference, the religious forms developed by such societies are particular to that group, and inextricable from other aspects of its cultural identity. In these circumstances, absolutized beliefs develop at the interpersonal level, associating religious experience solely with its relationship with our own group and its symbols. Such attitudes may be specific to the gods and ancestors of quite a small group, as in many of the groups of New Guinea researched by Diamond, but they may also expand into absolutized ethnic religions that influence a wide area.

Hinduism and Judaism are perhaps the best-known examples of thoroughly ethnic religions – but universal religions may also become heavily ethnicized (for example, the Narwari Buddhism of Nepal, in which monkship has become hereditary). Many of the instructions of the Torah from the Hebrew Bible are most easily comprehensible as ways of differentiating one tribe from another – but these instructions have become petrified and given absolute authority by association with Yahweh (who started out as a tribal God). For instance, the Israelites were allowed to eat any animal that had cloven feet and chewed the cud, but not any that did only one or the other.[12] Such rules are matters of authority, loyalty, and purity (see 2.e above), but not of universalized justice, care, or freedom. Such interpersonal absolutized commands gained their power from being associated with the Israelite experience of the God archetype, with striking epiphanies also being described in the Torah:

---

11   Diamond (2012) pp. 49–50.
12   Leviticus 11:1–4.

> Moses went up [Mount Sinai] with Aaron, Nadab, and Abihu, and seventy of the elders of Israel, and they saw the God of Israel. Under his feet there was, as it were, a pavement of sapphire, clear blue as the very heavens; but the Lord did not stretch out his hand against the leaders of Israel. They saw God, they ate and they drank. The Lord said to Moses, 'Come up to me on the mountain, stay there, and let me give you the stone tablets with the law and commandment I have written down.'[13]

These absolutized commands, of course, provide the shortcut beliefs that have mobilized the Israelites in tribal wars that have continued from the time of Joshua to this day. Is 'religion' to blame for such wars? Well, that all depends on whether you mean the religious experience or its projection, the meaning or the absolute belief.

The development, or possibly re-development, of ideological universalized judgements in ancient agricultural civilizations provides a further step onwards in the development both of provisional religion and of absolutized religion. The development of writing (discussed further in 3.e below), which first appeared in Sumeria at the end of the fourth millennium BCE, may have provided an encouragement to abstract consistency of judgement by 'freezing' propositions for constant reference – though this consistency was purchased at the expense of contextuality. As we see in the case above of the tablets of stone delivered to Moses, though, such consistency could also be readily applied to reinforce the authority of ethnic religious values prior to their universalization. However, a whole connected set of universalizing approaches were regulated by writing, and were subsequently subject to both provisional and absolutizing forms: trade and economic regulation, law, philosophy, administrative bureaucracy, and even embryonic science and technology. In this context, it was very likely that universalizing forms of religion would also arise, identifying religious experience with values that also required an expanding consistency of application beyond our own group.

The arising of ideological judgements in religion can be identified roughly with what Karl Jaspers famously called 'The Axial Age'.[14] Jaspers specifically identified universalizing modes of thought amongst key thinkers between the eighth and third centuries BCE in China, India, Persia, the Middle East, and Greece, including

---

13   Exodus 24:9–12.
14   Jaspers (1953).

Confucius, the Buddha, Zoroaster, and Socrates. In the case of Socrates, his form of questioning helped to propel those with unreflective interpersonal beliefs into greater awareness of their limitations, in an explicit search for universal beliefs.[15] I will return to this in 3.f below.

In the case of the Buddha, the Gangetic plain of India in his time was a place of increasing trade, prosperity, and intercultural contact, with a mixture of republican and monarchic forms of government. The Buddha was explicit about his acceptance of followers from all groups and classes. Uniquely, his Middle Way teaching not only promoted universality, but also awareness of the limitations of universalized ideologies as encountered in the form of the Buddha's religious teachers and ascetic peers.[16] With the Middle Way, the interindividual stage had been broached, and embodied provisionality[17] offered as a way of interpreting universalized beliefs. Not surprisingly, though, this was little recognized, and Buddhism was largely spread as a universal ideological belief accompanied by practices. Buddhist beliefs, like those of other universalized religions, got their power from their association with religious experience in the widest sense – whether this was the experience of meditative *jhana*, or just a glimpse of new potential realized only in devotion.

The ideological and interindividual levels of judgement potentially brought religion to the peak of its potential for supporting integrative human practice. However, its function as practical religion was very shortly hijacked by absolutization – not, this time, primarily of beliefs identified only with the group, but of universal beliefs interpreted as metaphysics. Metaphysics, as discussed in *Absolutization*, makes claims about 'reality' that are beyond any modification in the light of experience, and can thus only be held absolutely, whether they are negative or positive.[18] Absolutized religious beliefs readily show all the features of metaphysics, including foundationalism and circularity, infinite rationalization, claims of its own inevitability, and inflation of its practical significance.[19] For example, circularity is exhibited through the justification of dogma through circular reference to scriptures selected and interpreted by

---

15 I.5 d (pp. 151ff).
16 Ellis (2019) section 1.
17 Ibid.: 1.e, on embodiment and 3.c on provisionality.
18 I.4.a.
19 I.4.c–f.

traditional authority, and traditional authority justified through scriptures. Infinite rationalization is exhibited by the ways that God can be made compatible with any amount of evil, or an enlightened person can do any evil as 'skilful means' that are beyond our unenlightened understanding.[20] Claims of inevitability can be seen in the ways that religious beliefs are framed so as to make themselves unavoidable through their alleged essentiality – as in the supposed arguments for God's existence. Inflation of metaphysics is systematically seen in religion in the ways that 'belief' is credited with the effects that actually depend on the inspiration of archetypal meaning. I may judge differently on recollecting God, for instance, due to the effects of that association on the degree of contextualization in my mental state – not because my 'belief' in God makes the slightest difference to anything except its own absolutizing self-perpetuation.

These patterns in both the practical and dogmatic forms of universal religion have shaped the history of religion as we encounter it, both in the avowedly universal religions of Buddhism, Christianity, and Islam, and in the universalizing influence on ethnic religions. The latter can be seen, for instance, in the development of Vedanta from Vedic Hinduism, or in the move from henotheism (one god amongst many) to monotheism (one God alone) during the documentation of early Judaism found in the Hebrew Bible. At the same time, universal tendencies in religion have allowed people to overcome one kind of absolutizing judgement (in the process moving into ideological forms of thinking), but set up another kind – that which is systematized in theology, religious philosophy, and religious law.

Perhaps the most dramatic demonstration of the increased, but still limited, power of universalized religion can be seen in the rapid spread of Islam between the seventh and ninth centuries CE. The new ideological force of Islam united people who had previously been ethnically divided, using shared absolute beliefs, of both a religious and a political type. A new, disciplined, demand for consistency was focused on the Tawhid – the unity of God, from which the unity of his revelations, rules, and worshippers follows.[21] Whenever the Muslim armies conquered a new area, they could rapidly swell their ranks with new converts, at least partly motivated by the degree

---

20 See *The Skill in Means Sutra*: Tatz (1994).
21 Ellis (2022) 5.k.

of integrative consistency this demands. The unity of the caliphate crumbled in the ninth century, but a golden age of cultural creativity followed for several centuries after this, followed by the rise of the Ottoman Empire as a new claimed caliphate. However, from the fifteenth century onwards, the Islamic world went into a long, slow decline, gradually eclipsed by the West politically, economically, scientifically, and technologically. 'What went wrong?' asked historian of Islam Bernard Lewis in his book of that title. His answer, in the end, is primarily the loss of freedom in the Islamic world[22] – though it may be more a question of how freedom is interpreted. When the West adapts to new conditions with greater flexibility in the interpretation of values, as promoted for instance by democracy, economic competition, expanding higher education, and the liberation of women, this is more likely to create adaptiveness, just as the Islamic world has often rigidified – even though both are still working primarily at the ideological level of judgement.

Metaphysical claims are fragile,[23] as we see in the constantly shifting schisms and reformations of religious histories. The sixteenth-century Lutheran Reformation in the Christian Church, for instance, was largely just a flip from the absolute authority of the Church to the countervailing absolute authority of scripture, though it could be argued that the new power this gave to individual judgement[24] may have incidentally aided more people to move from interpersonal to ideological forms of judgement. Though a flip in religious belief may thus help precipitate a stage transition, whether it does so or not is dependent entirely on the readiness of the individual (in their cultural setting) who thus advances. It does not indicate that the flip in itself is necessarily an advance. That point can also be illustrated by the differing possible functions of conversion between Christianity and Buddhism: I have met Western Buddhists for whom conversion to a metaphysical understanding of Buddhism and rejection of Christianity provided a release from many interlinked dogmas and a leap forward into a more practical framework, but I have also met Asian Christian converts *from* Buddhism for whom release from Buddhist dogmas into the fresh thinking offered by Christian movements offered a similar step forward.

22   Lewis (2002) p. 177.
23   I.2.c.
24   Dumont (1985).

The fragility of metaphysical projections has led Christianity and Islam into a long history of fissiparousness – the tendency to split easily.[25] The same also applies to Buddhist tradition when it has become bogged down in metaphysics[26] or appropriated by political concerns.[27] The same tendency has also been taken over by politico-religious ideologies such as Marxism (Bolsheviks v Mensheviks, revolutionary v democratic socialists, Leninists v Maoists, and so on). At the same time, the mystical strands of universal religions have increasingly found commonality by sidelining metaphysical claims and focusing more on religious experience[28] (although when this in turn is rigidified in metaphysical monism, it continues to cause some degree of conflict).

In sum, then, religious archetypes offer vital sources of inspiration going back to the earliest history of humankind. Practical religion based on the cultivation of these sources of inspiration, associated with religious experience, have also been the source of great integrative benefit in overcoming conflict in humankind. The 'Axial Age' launched the more systemic development of such inspiration at the ideological and even interindividual level. However, this helpful function has also been hijacked by projection and absolutization from an early stage, probably going back into the paleolithic, with religious 'belief' being used as a shortcut to bind a group. This is a shortcut that may aid groups in the short term, but not sustainably in the context of wider conditions. This shortcut at first absolutized the values of groups, but developed a new dimension when it was used at the ideological level of judgement, from the development of ancient religions onwards. The history of world religions can illustrate both the spread of archetypal inspiration and the conflictual effects of the ways it has spread metaphysical belief. No one given religious tradition is immune to this hijacking, which is a general feature of human history. Even though reforms within a dualistic metaphysical framework may sometimes precipitate stage advances, they do not themselves offer satisfactory substitutes for the practical functioning of religion as inspiration.

---

25   II.4.e (pp. 154–6).
26   See any introduction to the competing philosophical schools of Buddhism.
27   See Ling (1979) for evidence that Buddhist tradition is far from immune from conflict, including war.
28   Ellis (2018) 7.b; (2022) pp. 242–6.

3.d. Desire, Exploitation, and Liberation

> *Summary*
>
> Desire is part of the human condition, but can be absolutized positively into addiction and exploitation, or negatively into asceticism and ideologies of liberation. Cultural conflict between addicted and ascetic groups creates reactive counter-dependence. Addicted societies are shaped by exploitative inequality, but absolutized ideologies of liberation in reaction produce ineffective revolutions that neglect the incremental psychological conditions needed for equality. Nevertheless, both ancient and modern societies also show evidence that effective reduction of addictive inequality is possible.

Desire – to continue to exist, to eat, to reproduce, and so on, is a basic effect of autocatalysis in organisms, as discussed back in 1.a. It is thus an aspect of human embodiment, motivating us in the very continuance of our lives, yet at the same time the absolutization of desire has been recognized as problematic from early times. Attitudes to desire, with the social, economic, and ecological effects of getting what we want, offer another major sphere in which we can try to chart the process of provisionality and absolutization in human history.

As discussed in *Absolutization*, absolutized desire is associated with beliefs about our goals, or about the background conditions for reaching them, that lack any wider context, so that the fulfilment of those desires becomes the whole story. In Buddhist tradition this absolutization of desire has been called craving (*tanha*) and its frustration into hatred and reliance on delusion long recognized.[1] It is a central insight of the Buddha's Middle Way, however, that merely trying to extirpate or reverse desire as a whole (asceticism) is incompatible with our embodied position: it is addressing craving or absolutized desire through integration that is crucial, not contradictorily trying to beat desire in accordance with our desire to beat desire.[2]

Absolutization of desire can not only lead to conflict within ourselves[3] (as in the case of addiction), but also to exploitation of others, as one class represses another, and thus gross inequality in society.

1   I.1.b.
2   Ellis (2019) 1.e.
3   I.5.a.

Reacting against either desire itself, or the exploitation it feeds, however, can lead to equally dogmatic uses of ideologies of liberation, that neglect the interdependence of freedom with other values. Provisionality and absolutization in relation to desire can be traced through history in both our attitudes towards desire and asceticism on the individual level, and towards property, exploitation, and liberation from exploitation at the social level. Exploitation of others, one might say, is a social expression of self-indulgence, and liberatory ideologies involve an ascetic purging for society that can be just as conflictual if unintegrated.[4]

The basic distinction between our left-hemisphere driven response to objects and our right-hemisphere driven response to a recognized other living thing (see 1.h above) constantly influences our judgements on these issues. Do we treat humans and animals (including ourselves) as living things, or as objects to be appropriated in the pursuit of our goals? Do we treat non-living things in a decontextualized way, or as integral parts of a complex system in which living things are also involved? These questions can only be answered in the particular circumstances of an individual at a particular time, but historically we can also ask a third one: how far do the cultural habits of our group in a particular historical era encourage us to either be ruthlessly instrumental or to recognize others as we pursue our desires?

Graeber and Wengrow's discussions of the contrasts between different neighbouring past societies in this regard are revealing, showing a tendency for different groups to distinguish themselves from one another in their responses to desire. If one group constantly reinforces over-indulgence fuelled by exploitation of others, for instance, it is a common pattern for a neighbouring group to react against this with asceticism, a higher degree of egalitarianism, and a culture of self-discipline. Graeber and Wengrow call this *schismogenesis*. As an example, they discuss the contrast between the Native American groups of the Northwest Coast and those of California. Those of the Northwest Coast owned slaves and were much more clearly hierarchized by class. The slave-owning classes went in for conspicuous consumption, in lavish *potlatch* ceremonies in which candlefish oil was ladled profligately into the fire. In contrast, the Californian groups cultivated modesty, hard work, and self-discipline, with

---

4   II.5.e.

gruelling initiation ceremonies.⁵ One could probably identify similar patterns in a whole set of paired neighbouring groups throughout history: Athens v Sparta, US Southerners v Northerners, or even the Capitalist world v the Communist one during the Cold War.

The 'self-indulgent' groups are led into one kind of reinforcing feedback loop by the accelerating and widening effects of what may have started off as small differences. The effectiveness of exploiting others (perhaps arising from the secondary exploitation of farming by hunter-gathering groups) makes such exploitation socially acceptable, and the hemispheric switch to instrumentality that it involves thus becomes habitual. The excess goods produced by such exploitation then create and reinforce the expectation of wealth and habits of over-consumption, locking in dependence on exploited labour. A whole set of social arrangements and reinforcing rituals are organized around this expectation.

In contrast, the 'ascetic' groups maintain a higher degree of mutual recognition of each other as living and as persons, and are thus likely to be disgusted by the instrumentality and waste of the self-indulgent group. They may have a more integrated and more sustainable economy because of this, although they lack the short-term advantages that can be gained from exploitative labour. They are, however, locked into another kind of reinforcing feedback loop – that of repressing desire and deferring gratification. Max Weber famously discussed this pattern of deferred gratification in the development of early capitalism out of Protestant Christianity, with the constantly-postponed goal being equally either a reward in heaven for belief or a reward of God-given wealth on earth for investment.⁶ The conflicts of this ascetic approach are more likely to be internalized, but are nonetheless conflicts, and they are heavily reliant on metaphysical belief in some kind of cosmic justice system that ensures the equivalence of later payouts.⁷ They are also more likely to produce rigid rule-following, creating conflicts over the application of rules.

The riddles of alcohol consumption provide a further instance of this contrast. Why do humans consume alcohol, despite its toxicity, not just in small quantities but in outbursts of excessive consumption? The most coherent answer to this seems to be that we use it

---

5   Graeber & Wengrow (2021) pp. 199–204.
6   Weber (1930) chs. 4 & 5; also Ellis (2001) 3.f.x.
7   VI.6; iv.4.g.

as a way of displaying our strength: if we can weaken ourselves unnecessarily in a particular context of display, it shows that we have so much strength that we can even afford to weaken ourselves and still be strong[8] – we can 'take our drink'. This is clearly a function of the self-indulgent cultural mode, and its relationship to reinforcing feedback loops can readily be seen: socially sanctioned over-indulgence is self-fuelling and self-accelerating, and so is the process of addiction at the level of the individual body, with these two spirals further fuelling each other. The ascetic mode is, in contrast, likely to emphasize the virtues of abstention to avoid all the negative health-related, social, economic, and intellectual effects of alcoholic indulgence. Mere denial of the urge to drink alcohol without a wider alternative context, however, puts a heavy reliance on a merely repressive rule, and perhaps on fragile absolute beliefs that are used to justify that rule. Schismogenesis can be seen, for instance, in the prohibition of alcohol in Islam,[9] or the development of prohibition in the early twentieth century US – each in reaction to earlier self-indulgent cultures and dependent on absolutized religious rules.

Buddhist tradition classically articulates the two extremes that the Middle Way avoids in terms of self-indulgent as opposed to ascetic attitudes to desire – as 'eternalism' (*sassatavada*) v 'nihilism' (*ucchedavada*). These two extremes were signified by the Forest and the Palace in the story of the Buddha's early life, with each being assumed dependent on clusters of metaphysical beliefs: realism, moral absolutism, freewill, and cosmic justice in the case of eternalism and idealism; moral relativism, determinism, and the denial of cosmic justice in the case of nihilism.[10] The period of the Buddha's early life where he engages with the absolute beliefs of the Forest culture has two episodes – first the encounter with spiritual teachers (who are found to be limited in their understanding), and then the encounter with a group of ascetics who believe that self-mortifying practices can lead to future reward under the cosmic justice system of karma. I have previously explored this story in depth as symbolic

---

8   Diamond (1991) ch. 11.
9   Michalak & Trocki (2006).
10   In Ellis (2019) 4.d I explain why such assumed clustering does not offer an adequate account of the extremes avoided by the Middle Way: the clusters are mutually supportive but contingent.

of the process of finding the Middle Way in individual experience,[11] but we could also see it as a process of cultural encounter. The Palace and the Forest, associated with self-indulgence and asceticism respectively, are cultures in schismogenetic relationship with each other. In the context of wider Indian culture, they may be related to the influence of different cultural groups that have moulded Indian tradition in different ways – the 'Aryan' influence of the warrior culture being associated with exploitation and self-indulgence, whilst the 'Dravidian' culture was associated with individual ascetic practice.[12] It may well be the case, then, that the Buddha's Middle Way offered some resources for the possible reconciliation of two cultures, as well as the addressing of conflict in individual experience.[13]

The feedback loops of self-indulgence on an individual level constantly interact with the social and cultural conditions, because of course one cannot be self-indulgent unless the environment is providing one with the resources to do so. As long it keeps doing so, though, various reinforcing feedback loops interact: most basically there is the dopamine loop of satisfaction at getting something one wants, then there is the constant reinforcement of our beliefs about the world (and about the values we should apply to it) linked to that dopamine loop in a wider loop. In the wider environment, though, we need to keep making sure the resources are present to feed our self-indulgence, which means that when stressed by the possibility of our satisfaction being withdrawn, we are likely to take short-cuts to make sure these continue, by absolutizing the values and background beliefs. For instance, if we enslave others to feed our self-indulgence, we have to constantly reinforce our instrumentalization of them and repress our awareness of their fellow humanity. The ensuing institution of slavery at a socio-political level then entrenches these absolutizing assumptions, making it far easier for an entire culture to keep making the same absolute assumptions as the basis of slavery.

---

11 Ellis (2019) section 1.
12 For a classic (and now definitely out of date) articulation of the historical theory behind this see Dutt (1970). The racial labelling of the two sides may well be an over-simplification, but the two tendencies nevertheless appear to be the basis of discernible cultures.
13 See Ellis (2019) 2.e.

Gross inequality in society, whether of formal chattel slavery or of some more contractually-based form of exploitation, will then keep getting culturally reinforced through a number of other reinforcing feedback loop mechanisms: conditioning in both a social and epigenetic form to accept one's class, *deformation professionelle* (the embodied constraint of both bodies and minds to a particular range of tasks), technology further embedding that deformation, money and its investment adding to the mounting wealth and power of the privileged, and institutions of inheritance further increasing the gap as generations progress. If unchecked by any other social or political mechanisms, social inequality clearly has a systemic tendency to accelerate, just like other reinforcing feedback loops we have seen. There are, of course, ways it can be checked: in the modern world by graded taxation, universal education, and the welfare state. However, if it continues unchecked, increasing fragility and eventual social collapse is likely: whether that collapse is precipitated by environmental degradation from the accelerating despoilation of resources, or from the social reaction of workers who, in the famous words of Marx and Engels, 'have nothing to lose but their chains'.

The revolutionary reversal of highly unequal societies is evidently not just a relatively modern phenomenon. We may think first of the French and Russian revolutions, but Graeber and Wengrow present evidence of a likely revolution in the Chinese city of Taosi around 2000 BCE.[14] Whilst the original excavators thought that this had led to a 'state of anarchy', Graeber and Wengrow argue that since it lasted a couple of centuries, and resulted in an apparently prosperous redistribution of wealth (as evident in the excavated houses), this may well have been simply the reversal of a rigid class hierarchy back to a more egalitarian system. Perhaps at that earlier stage of human history, when fewer turns of the wheel had entrenched inequality so deeply, it was easier to achieve this in a sustainable way than it was later. However, the evidence of modern revolutions is not so promising. As is well-known, the French Revolution of 1789 resulted, not in a stable egalitarian society, but in the Terror and then the imperialist rule of Napoleon. The Russian Revolution of 1917, again, resulted quite quickly in authoritarian rule, in the widespread intellectual repression of any alternative beliefs to

---

14 Graeber & Wengrow (2021) pp. 325–6.

Communism, and in the longer term, in an increasingly unequal society.

We don't need to look far in systems dynamics to understand why revolutions generally fail. Systems are constantly changing, but their changes are unavoidably incremental, even if at times they can reach relatively rapid tipping points.[15] Systems, particularly complex ones, are not capable of changing instantaneously in response to the imposition of an ideological model. In human societies, a whole set of entrenched conditions make it very difficult to immediately halt exploitation: as already mentioned above, those conditions include social and epigenetic pressures, *deformation professionelle,* and the economic effects of wealth and its constant recirculation. In particular, those who concentrate on change at the socio-political level often completely neglect to consider the mass psychological changes that are necessary to stabilize those socio-political changes. You cannot liberate people who are so much in the habit of subservience, so habitually stressed, or so obsessed with certain distracting dopamine-fuelled satisfactions, that they will immediately click back into the same old shortcut patterns of judgement as soon as any kind of complexity appears.

The ideology of liberation may thus have a crucial role in expanding the *options* that people consider, but its success in changing their whole spread of beliefs and practices depends on its degree of sustained engagement with the complexity of a whole range of conditions. Such sustained engagement becomes more likely if a weight of the population sufficiently recognizes the range of socio-political values that are all interdependent in the maintenance of human society. As already discussed in 2.e, these values, as identified by Jonathan Haidt, can be seen as consisting of loyalty, authority, purity, justice, care, and freedom. The interpersonal stage tends to prioritize the first three, namely group-oriented values, whilst subordinating the other three (universal values) to the first three. However, emergence into the ideological stage can have the helpful effect of prioritizing the three universal values over the group-orientated ones. As also discussed in 2.e, this seems to be the key condition that has enabled Western democracies in the twentieth century to rediscover a degree of egalitarianism (perhaps most marked in the Scandinavian countries, with their high levels of

---

15 II.3.a & b.

education[16] and low levels of inequality[17]). However, they reached this position by a whole set of gradual adjustments – including the development of democracy itself, bureaucracy, higher education, and freedom of speech, aided by sufficient economic prosperity.

'The ideology of liberation' may obviously describe a range of ideological motives for challenging an unequal society, but the key point offered by the historical evidence seems to be that liberation has both a provisional and an absolutized form, with the former rather than the latter required for the successful braking of the reinforcing feedback loops associated with inequality. Absolutized forms of liberation will not only prioritize the universal values of justice, care, and freedom over the group-orientated values, but also exclusively focus on one of those three values over the others, or one *interpretation* of those values over the others, resulting in a lack of contextual adjustment in the application of the value.

In the first case, universal values that completely neglect the group-based values are in danger of absolutization, as people react against what seem to them abstract bureaucratic requirements, and fall back on the more immediate values offered by loyalty to their group or of charismatic leadership. It is not surprising that revolutions built on these selective, non-systemic foundations have crumbled.

In the second case, if, for example, if we fixate on justice as equality and interpret equality in terms of equal incomes, we may try to transform the whole economic system to bring about this end. In that process we neglect the effectiveness of economic activity, the side-effects of the bureaucratic methods needed to impose it, and the value of individual freedom to allow innovation – as seen in the failures of Soviet Communism. If, alternatively, we fixate on freedom and adopt a libertarian approach that excludes considerations of justice, we will unavoidably neglect the question of whose freedom we are talking about, boosting the resources and the capacity to act of those who already have those things to some degree, but doing little for those that are not in a position to take advantage

---

16 https://ourworldindata.org/global-education#years-of-schooling (accessed 2022).

17 As measured by the Gini coefficient: https://data.worldbank.org/indicator/SI.POV.GINI (accessed 2022).

of freedom. Widening inequality in the US,[18] for instance, follows the increasing political influence of libertarian narratives on the Republican Party since the 1980s.[19]

If we put a highly selective absolutization of values into operation without the rest of the socio-political foundational values, practical fragility results. However, perhaps a more common outcome is that these selective values are never actually put into operation: instead, their role is only to offer an ideological substitute that binds the group, but that then may be used in a way that practically produces the reverse of the value, with freedom being used to justify oppression and justice used to justify restriction. Graeber and Wengrow offer insights on this by discussing the way that the idea of equality can offer 'a system of equivalence' that allows all citizens to be treated in the same way regardless of actual differences in circumstances or needs.[20] Thus, for instance, debts are never forgiven because everyone must 'equally' pay their way (regardless of the injustices of the system that created the need for debt), and undischarged debtors are enslaved. Authoritarian rule supported by implacable bureaucracy is needed to put this into operation, but such authoritarian rule can ironically appeal to the justice of a completely abstracted and decontextualized appeal to equality to create extreme inequality.

In contrast, many ancient societies had systems of regular debt forgiveness or other economic equalization, including the ancient Israelites: the book of Leviticus lays down that every fiftieth year should be a 'Year of Jubilee' in which everyone returns to their original property holdings at the previous year of Jubilee, regardless of financial transactions between.[21] The Andean region at the time of the Incas (though not due to the Incas themselves) had an even more radical system: debts were forgiven every year.[22] However 'unjust' debt forgiveness might seem to a rigid left-hemisphere rule-following view, it had the effect of cutting rounds of reinforcing feedback loops that would otherwise build up, fuelled by the inequalities that were created by debt. Such systems

---

18 Between 1974 and 2019 the US Gini coefficient rose from 38 to 41 (World Bank, ibid.)
19 Wisman & Smith (2011).
20 Graeber & Wengrow (2021) p. 425.
21 Leviticus 25:13–55.
22 Graeber & Wengrow (2021) p .423.

are not the result of ideologies of liberation, but rather of a more adequate complexity modifying an existing system and making it more sustainable. Although the Israelites made them part of the absolutized laws of Yahweh, there seems to be evidence of a regular Babylonian decree of *nisharum* on which the year of Jubilee may have been modelled, making the Year of Jubilee part of a wider adaptive cultural pattern.[23]

There are many questions about the practicality of debt-forgiveness laws, but they are, at least, a positive historical indication that the system can be framed differently to include balancing feedback. More broadly, the success of some countries (and indeed to some extent the world as a whole) in somewhat reducing inequality[24] (for instance through targeted development and welfare), is also indicative that ideologies of liberation do not have to be reactive and fragile, but instead can develop integrative forms that succeed in addressing all the major values. In the process, they will have managed to incorporate enough balancing feedback loops into socio-political judgement to create a more sustainable decision-making process, rather than continuing on the path of trying to impose a heavily left-hemisphere orientated ideological scheme on complex conditions, regardless of the feedback.

The type of politics that can succeed in this way will need much more detailed discussion in a later volume of this series,[25] but in general it must be one that follows the four criteria discussed in the later part of *Absolutization*: namely being practical, universal in aspiration, judgement-focused, and error-focused. This does not necessarily mean the adoption of one existing ideology (as long as that ideology is not entirely confined to interpersonal values) so much as care in the interpretation of that ideology. Whatever values we have in the foreground, we need to remain aware of others in the background to keep that ideology provisional: a requirement that can only in part be addressed by bureaucratic or legal checks, but also depends on moves towards an interindividual or Middle Way style of judgement.

---

23 Westbrook (2016) pp. 215–21.
24 https://ourworldindata.org/income-inequality#within-country-inequality-in-rich-countries (accessed 2022).
25 IX.

3.e. Literacy and Idolatry

> *Summary*
>
> The development of writing has facilitated both contextualized and abstracted representation, and the latter, due to re-thematization when learning to read, seems to be understood in a way that can be easily absolutized. Both the power and the absolutizing danger of abstracted representation were reflected in the monotheistic prohibition of the idolatry of images, but which was then not extended to the written word. Instead, iconoclasm has created new conflicts between interpersonal thinking, and ideological thinking dependent on literalized written propositions.

Writing first seems to have emerged in Sumeria around the end of the fourth millennium BCE, and again independently in China by 1200 BCE and in Mesoamerica by around 500 BCE. Whilst writing seems to have developed at first for purely practical purposes, like recording the numbers of livestock or wheatsheaves,[1] these practical purposes involved contextualized representation: for instance, three lines on a clay tablet was recognized in a particular context as representing three wheatsheaves. The interesting feature of writing and its effects for the history of the Middle Way lies in the way that *contextualized* representation, for specific purposes, could change to *abstracted* representation without such a context. Whilst contextualized representation is still provisional and open to modification in the light of experience, abstracted representation is not, because no new information from the context is feeding into it any longer. Of course, abstraction was possible without writing, but writing seems to contribute particularly to the process by removing a supposed representation from its context. A written record might not only form a record of a trade deal long after it was made, but also record the supposed timeless truths uttered by a god. Once again, then, a balancing feedback loop, whereby new information was communicated across time and space to help us adapt more fully, was hijacked into an instrument of absolutization, creating and reinforcing feedback loops of ideological certainty.

The distinction between contextualized and abstracted representation can be seen in part in the evolution of writing systems themselves. The earliest elements of writing were *glyphs*, markings,

1   Schmandt-Besserat (2006).

the meaning of which are heavily reliant on the context. A series of scores on a clay tablet representing a number of sheep could easily represent something else in a different context. Glyphs are turned into *graphemes* when their relationship with an idea or a sound becomes more stable – for instance that a simplified and schematized picture of an ox becomes a character for an ox. The grapheme may then go on to develop phonetic associations and then a phonetic meaning: as in the case of 'A' representing a vowel sound, although in its capital roman form it still bears the schematized horns of the ox (the word for which originally began with that sound). This stability across contexts, though, begins a process of abstraction. The 'A' character starts to mean any ox in any context, and then any of a range of associated vowel sounds – again regardless of context.

Graphemes formed of combined letters in a phonetic script can then constitute words. Those written words give us a new freedom to imagine, either through their associations in isolation or as representational sentences describing possible states of affairs. It is this freedom to imagine that enabled literary uses of writing from an early stage: the earliest known literary text being *The Epic of Gilgamesh*, the earliest surviving versions of which go back to early in the second millennium BCE.[2] Although this epic may well have existed in oral form before being written down, its transmission in writing enables new possibilities to be considered by new people over time who would not have encountered the oral version. The very stability of a provisional world, say in a poem or a story, allows us to share it as an aspect of our social awareness, and take it more seriously as a possibility. The degree of abstraction enabled by writing could thus enable imagination for greater provisionality, as well as provisional beliefs in practical contexts.

However, it is the ways that writing has encouraged absolutization that are perhaps more striking. It is Piaget who provided the psychological resources for understanding the key process involved in literacy that may lead us to regard written texts differently from the spoken word. As we learn to understand spoken language first, we *thematize* by learning to associate the different parts of a proposition with the supposed representational meaning of the whole, using our implicit understanding of grammar. If we did not do this, language would just be a set of associative fragments, rather than

---

2   Sandars (1960).

a helpful tool for understanding our experience and communicating for practical purposes. We might associate meaning with 'John', 'loves', and 'Mary', but have no idea what John and Mary have to do with each other. When we learn to thematize, though, we apply an implicit understanding of grammar: using the conventions of English word order, 'John loves Mary' has John as the subject and Mary as the object of his love (not necessarily the other way round!). However, when we learn to read, we have to go through the whole process of learning to thematize again from the beginning, by associating the relationships between *written* words with the representation in a *written* proposition.[3] The meaning of the written proposition thus seems different from that of an equivalent spoken one. This may help to explain the observable ways that people often treat the authority of written texts differently from oral utterances.[4] Both writer and reader (or auditor) tend to treat the text in a more decontextualized way, which often means attributing authority to it that is absolutized, because it stems from beyond the situation where it is applied (thus also preventing it from being questioned or modified within the practical terms of that situation). That absolutization can also create a flip by which the text as a whole is rejected, without any contextual consideration of its degree of merit.

The Torah from the Hebrew Bible records some of the practical consequences of this, in the form of the injunctions to follow written rules that are given absolute status over largely pre-literate people. Whether or not Moses's acquisition of the law of Yahweh written on tablets of stone from Mount Sinai has any historical basis, the story records an attitude to the written word that indicates its absolutized use. By comparison to the use of images (which are individually interpreted using the imagination), these written words provide reliability and universality that is taken to be immune from all such 'subjective' interpretation. The use of images, by contrast to the written word, is 'idolatry', and is prohibited.[5] Although at this stage Yahweh is not yet a universal God, but still a tribal one, one of the key conditions for universalized metaphysical religion is laid

---

3   Perfetti & Goldman (1975); Ferreiro (1985); also discussed in Ellis (2022) p. 220.
4   Fondacaro & Higgins (1985).
5   Exodus 20:3–4; Leviticus 26:1.

down. The same prohibition of idolatry (*shirk*) later became a key tenet of Islam.[6]

The prohibition of idolatry (as normally interpreted) incorporates the insight that the God archetype is not a specific representation, but an experience that is not represented.[7] This is an insight reflected in Islam by the idea that the sin of *shirk* involves 'giving God partners', or confusing God with other things.[8] In its most common interpretation in relation to visual images, this offers a recognition that absolutizing an image deprives us of the more universalizing context that the written word might offer. The 'idolater' is assumed to project a belief in supernatural 'existence' onto the visual image itself, ignoring the wider context of our own responsibility in worshipping an image. Of course, the worshipper of a visual image may instead merely be prompted into archetypal inspiration by that image, which is associated with religious experiences and wider values. However, such a response requires another level of reflection that the first prohibitors of idolatry perhaps judged more likely if we are thinking at an ideological level, rather than responding more directly either to the image as an instrument of our wishes or as an expression of our group identity.

Literacy alone, of course, is neither necessary nor sufficient for ensuring judgement at the ideological level, but literacy and an ideological level of judgement may still be seen as closely linked in societies that have literacy and come to depend on it as a basis of education. This association is particularly strong in the monotheistic cultures, where *consistency* of written rule-following across contexts is prized as the implication of one universal God. The constant reviewing of written propositions that stay still, rather than being subject to the constant contextual changes of our circumstances, is the basis of the legalism that both the Torah and Islamic law require, so it is hardly surprising that they are suspicious of more powerful and transient forms of inspiration.

A consistent application of the insights behind the prohibition of idolatry, however, would lead us to apply it just as readily to the absolutized written word as to visual images. The written word, too, leaves us in danger of mistaking a specific attempt at representation for the thing being represented, a particular description or

6   Discussed in more detail in Ellis (2022) pp. 249–50.
7   I.5.b (p. 142); also Ellis (2022) 5.f & k.
8   Qur'an 4.36; Murata & Chittick (2000) p. 49.

prescription for an absolute one. Sceptical argument, which I will consider in the next chapter, makes this clear, by pointing out the gap between embodied experience and representation. Historically, though, this point seems to have been at best thought implicitly by a few mystics until relatively recently. In the twentieth century, at the time of the Second World War, it was articulated by Simone Weil:

> Idolatry comes from the fact that, while thirsting for absolute good, we do not possess the power of supernatural attention and we have not the patience to allow it to develop.[9]

> Idolatry is the name of the error which attributes a sacred character to the collectivity; and it is the commonest of crimes, at all times, at all places.[10]

The 'collectivity' here, I would interpret not to mean only the appeal to interpersonal values, but also the shared and socially-reinforced ideological values one may develop in the ideological stage in reaction to interpersonal values. Such ideological representations do indeed 'partner' God, giving infinite justifications to finite statements.[11] Idolatry, then, is more broadly another term for absolutization.

The unrecognised idolatry of the written word, coupled with militancy in the rejection of visual images, is an excellent example of the hijacking of a provisional religious insight, so that it became subject to the fragility and conflict of absolute religious belief. The history of this begins in the episode of the golden calf narrated in Exodus – when in Moses's absence the Israelites created a golden calf to worship, and were punished by the extermination of 3000 of them.[12] It continued with the struggles between iconoclasts and iconophiles in the eighth-century Byzantine Empire,[13] and the opposition to Catholic images in the Protestant Reformation. As Iain McGilchrist comments, the latter 'attempted to do away with the visual image, the vehicle *par excellence* of the right hemisphere, particularly in its mythical and metaphoric function, in favour of the word, the stronghold of the left hemisphere, in pursuit of unambiguous certainty.'[14]

9   Weil (2002) p. 60.
10 Weil (2005) p. 76.
11 I.4.d.
12 Exodus 32.
13 Treadgold (1997) pp. 150–3.
14 McGilchrist (2009) p. 315.

Iconoclasts and iconophiles are a further example of cultural groups that have played off each other in a reinforcing feedback loop of mutual reaction – the schismogenesis recognized by Graeber and Wengrow. Catholics and Protestants in Europe, Muslims and Christian Orthodox in the Near East, Sunni and Shi'ite Muslims in the Middle East, and Muslims and Hindus in India offer further examples of such schismogenetic pairs. Perhaps the most recent striking example of this was the destruction of the Bamiyan Buddhas in Afghanistan by the Taliban in 2001, an act drawing on the Islamic doctrine of *shirk*, but clearly also a symbolic rejection of wider 'Western values'.[15]

Iconoclasm in such contexts can also become associated with asceticism in one's approach to desire, fed by the idea that images provide sensual enjoyment which feeds a cycle of craving. This absolutizes an insight that requires critical differentiation. Images can indeed feed obsessive cycles as well as unconscious inclinations (particularly in modern advertising and propaganda facilitated by communications technology – something I will return to in 3.h), but in these cases, images are being used to narrow contextual awareness rather than to extend it, when images can have both of these functions. 'Worship' of religious images, like sublime appreciation of works of art, extends the context of meaning for our judgement rather than determining it. In some ways, then, the narrow rejection of images is inseparable from the repression of human desire in general, and has similarly self-destructive long-term effects. The standpoint of representationalist absolutization of the written word, from which images can be rejected, similarly can have short-term value in particular contexts, but in the longer-term constricts the range of conditions we can respond to.

---

15   Reza Husseini (2012).

3.f. Sceptical Philosophy and its Appropriation

> *Summary*
>
> Sceptical argument implies even-handed and consistent questioning to help us recognize uncertainty and thus consider new possible beliefs. It seems to have originated in Indian Buddhism, from which it spread to Greece in the form of Pyrrhonism. However, even in these ancient sources scepticism is not consistently applied, and its subsequent history is one of constant defensive misinterpretation, through the techniques of appropriation and lumping.

Sceptical philosophy is the practice of asking questions that remind us of general uncertainty. In *The Five Principles*, I discussed in detail the patterns of sceptical argument, the ways it is often misunderstood as negative or impractical when it is neither, and the importance of applying these arguments even-handedly rather than selectively.[1] At that point I deliberately avoided any discussion of the historical origins of sceptical argument. However, the way that sceptical argument first came about, and was then thoroughly misunderstood and misappropriated, is a historical matter best understood within the general patterns I am exploring in this section of the book. Once again, it is an adaptation of provisional value, enabling balancing feedback, which was hijacked in the service of absolutization and reinforcing feedback.

It is impossible to tell when and how sceptical argument first arose. Probably long back into the paleolithic, people were capable of considering the possibility that things they previously believed, or that others believed, were not necessarily correct – that is, of doubting. Sceptical argument as such, though, needs to be generalized, which makes it unlikely to be developed prior to the ideological level of judgement, and probably much aided by the general reflectivity that the spread of writing eventually helped to support, even if the first sceptical discussions were entirely oral. To question *all* claims, questioners must be capable of doubting authority, which gives the capacity for sceptical questioning an interdependent relationship with the ideology of liberation.

The earliest recorded sceptical arguments are found in India, both in the Buddha's teachings and in the arguments of some of

1   II.1.

his interlocutors. In the Buddha's responses to the questions of Malunkyaputta, who wanted certainty on metaphysical claims such as whether the universe is infinite or eternal, the Buddha remained silent. When pushed he then gave the famous parable of the arrow, in which he compares the asking of such ultimate questions to the obsessions of a man pierced by an arrow who wanted to know all about the origins of the arrow before he pulled it out.[2] The Buddha's answer clearly identifies the *practical* problems created by absolutization, showing that his motive for not engaging in metaphysical questions was the wish to try to avoid the proliferation of reinforcing feedback loops they were likely to lead us into.[3] There is thus no reason to assume (as the Buddhist tradition unfortunately often does) that the Buddha's 'silence' on metaphysical questions is confined only to certain metaphysical issues that happened to be mentioned in the texts, rather than being a practical response to metaphysics in general,[4] or that the Buddha was withholding 'knowledge' of the correct answer to such questions that was only comprehensible to him. On the contrary, the Buddha's recognition of the drawbacks of absolute beliefs was consistent with a recognition of the limitations of the embodied human situation: a point we can see him recognizing, more than anything, in his discovery of the Middle Way as a point that addresses human uncertainty between the false certainties of absolutization on both sides.[5] Since this uncertainty between opposed false certainties is the basis of sceptical argument, the Buddha was clearly a sceptic.

However, Buddhist tradition has tried to distance itself from scepticism, perhaps realizing that an application of scepticism to dogmas such as those of karma and rebirth could undermine the absolute beliefs that had become associated with the tradition. One way in which it has done that can be found in its attack on Sanjaya Bellatthaputta in the Pali Canon. Sanjaya had a very similar approach to the Buddha's, if we judge from the presentation of his view in the Pali Canon.[6] He merely makes explicit the even-handed refusal to engage in metaphysical polarizations that the Buddha routinely presented in silence. However, for some reason Sanjaya

2 Majjhima Nikaya 63 (Ñanamoli & Bodhi 1995 pp. 533–6).
3 Ellis (2019) 3.d.
4 Ibid. 4.c.
5 Ibid. 1.e & f.
6 Digha Nikaya 2:31–3 (Walshe 1995 p. 97).

is presented as 'stupid and confused'. Of course, we don't have any information on how Sanjaya interpreted the practical implications of metaphysical agnosticism, but for Buddhists to attack that agnosticism itself seems to show that from a very early stage they were already moving away from the insights of even-handed sceptical argument, and into the hypocrisies of merely selective scepticism.[7]

The scepticism of the Buddha's Middle Way is reflected later in Buddhist tradition in the doctrine of Emptiness of the Perfection of Wisdom literature: as philosophically formulated, for instance, by Nagarjuna. This attempts to give a philosophical formalization to the insights of uncertainty and even-handedness in the Buddha's Middle Way. Nagarjuna's *Mulamadhyamakakarika*, for instance, works systematically through a range of different types of belief, asserting the even-handed emptiness of all of them, including beliefs about nirvana.[8] Again, one could only claim that this was not scepticism by misunderstanding the implications of sceptical argument in the first place, and assuming that 'scepticism' necessarily implies something negative, selective, or impractical. The scepticism of the Mahayana is also integrated into its spiritual practices, with Emptiness, for instance, being an object of insight meditation. However, the wider socio-political context of Mahayana Buddhism contains other sources of reinforcing feedback loops that may conflict with the sceptical insight and limit its application, such as the absolute authority of gurus, which was in turn used to justify theocratic government in Tibet. A scepticism that cannot be applied to political matters is not a fully generalized scepticism.

The links between Indian and Greek scepticism have become well-established,[9] though exactly what form they took is much more a matter for speculation. It is reported in Diogenes Laertius that Pyrrho of Elis, the founder of Greek scepticism, travelled to India with Alexander's armies in the fourth century BCE, and was there influenced by the Greek 'gymnosophists'. It was thus long thought that early Buddhism influenced Greek philosophy. However, Christopher Beckwith, in his book *Greek Buddha*,[10] offers a radically different historical picture, which he arrives at by relying only on datable sources, and interpreting the Pali Canon largely

7   Ellis (2019) 4.a.
8   Nagarjuna (1995).
9   Kuzminski (2008) ch. 2.
10 Beckwith (2015).

as a retrospective rationalization of a later and more dogmatic Buddhism. Beckwith concludes that the Buddha was a sceptical philosopher originating, not in the Ganges Valley, but in Scythia (now Ukraine), and that early Buddhism grew up in Gandhara (northwest India) in reaction to Zoroastrianism. He finds no evidence that the Buddha taught the Four Noble Truths or 'enlightenment'. Instead, Pyrrho's scepticism, a new departure for Greek philosophy, is seen as a direct importation of a new movement that was already Greek-influenced. Beckwith's approach, if nothing else, at least offers an antidote to over-attachment to specific historical claims about Buddhist, or indeed Greek, sceptical traditions, showing what radically different interpretations of the evidence can be rigorously supported. If Beckwith is correct, of course, then the origins of Buddhism may be much more provisional than the Buddhist tradition depicts them as being.

Fully developed sceptical argument arrived in the context of Greek philosophy with Pyrrhonism, a tradition now known largely from its later writings in the Roman period (especially those of Sextus Empiricus), with its earlier Greek ones only available in fragments. What is reasonably clear is that Pyrrhonism was linked to practice, and was even-handed in its interpretation of sceptical arguments. It was also more consistently critical of metaphysics than the Buddhist tradition became. However, its practical judgements depended on a distinction between what was 'evident' and 'not evident'. This ignores all the ways that we direct our attention in response to how meaningful we find what we sense, and the ways we are thus at least partially responsible for what we find 'evident' in our experience.[11] This leads us straight into further dogmas, such as epistemic determinism and moral conventionalism, rather than offering a model of provisionality. In my assessment it is thus mainly for its critical perspectives (its sceptical argument) that Pyrrhonism is valuable, rather than its positive proposals for Middle Way practice.

It's also arguable that some elements of sceptical practice were introduced into Greek philosophy before Pyrrho, by the figure of Socrates, although for our understanding of Socrates's method we are almost entirely dependent on Plato's depiction of him. Socrates has become well-known for his 'Socratic questioning', which

---

11 For more detail on these problems with Pyrrhonist doctrine, see my review of Douglas Bates's book *Pyrrho's Way* at https://www.middlewaysociety.org/books-philosophy-books-pyrrhos-way-by-douglas-c-bates/

continually asked his interlocutors for further justification of their claims, and for his use of dialectical methods. However, the goal of those methods seemed to be the isolation of essential truths to be separated from falsehood, rather than a general recognition of uncertainty. Although he was well-known for acknowledging the extent of his own ignorance,[12] this seems more likely to have been a matter of personal epistemic humility rather than a general recognition of uncertainty. Plato's use of the figure of Socrates in his dialogues rather illustrates the beginning of the *appropriation* and *lumping* of scepticism that has continued amongst philosophers and others from ancient Greece and India through to the present day.

Appropriation and lumping are two kinds of common absolutizing defences against agnosticism which I discussed in *The Five Principles*.[13] In appropriation, proponents of an absolutizing view claim that an agnostic view actually supports their view, whilst in lumping they claim that it is actually part of the opposing view and can thus be dismissed. The appropriation of scepticism uses it selectively to try to justify some absolute claim, whilst ignoring the way that its general application would undermine that absolute claim. The lumping of scepticism treats the even-handed recognition of uncertainty as a type of denial, and thus rejects it along with those who deny the lumper's absolute position. Both of these have become routine responses to sceptical argument throughout history, to such an extent that the very possibility of even-handed agnosticism is usually forgotten or denied. I only have space to offer a few examples.

There was a long medieval gap in Western philosophical history during which sceptical argument was effectively forgotten, and criticism of any kind was highly localized within an absolutized framework that was taken for granted. It was the rediscovery of Sextus Empiricus during the sixteenth and seventeenth centuries that seems to have prompted new responses to it.[14] The earliest of these were highly selective: Erasmus, for instance, concluded that sceptical questions merely led one back to the certainties of Catholic dogma,[15] whilst Montaigne, although he made extensive use of sceptical insights in his personal reflections, likewise refused to apply

12 E.g. *Apology* 21d.
13 II.4.d.
14 Popkin (1964).
15 Ibid. ch. 1.

them to ethical or political issues, where he fell back on the conventional certainties of his day.[16] The *locus classicus* of the appropriative response, however, is Descartes. In his highly influential *Meditations* of 1641,[17] Descartes used sceptical argument to try to reveal certainty, with the idea that if one used it as far as possible one would peel away error and reveal the truth. Descartes thus set out to doubt everything, but reached what he felt must be self-evident certainty beneath. What he conceived clearly and distinctly, he claimed, must be true – an *a priori* certainty known through reason alone. Being apparently certain of his own existence, he then reasoned 'clearly and distinctly' back up from this foundation to a whole set of other metaphysical beliefs, including the ultimate reality of God and of the world. Perhaps Descartes simply lacked the awareness that his own self-experience was constantly changing, let alone the psychological insight that mere clarity is no guarantee of truth. At any rate, his work was so influential that even modern philosophers who disagree with him still talk about 'Cartesian scepticism' – as though Descartes was remotely sceptical!

The other absolutizing defence mechanism, that of lumping, is best illustrated by the somewhat later responses of Hume during the eighteenth century.[18] Hume recognized that sceptical argument posed a systematic challenge to all claims of 'knowledge', and faced up to it seriously – perhaps the first post-Renaissance figure to do so. However, Hume also made negative assumptions about the implications of scepticism, concluding that it was impractical, and that 'nature' drives us to dogmatic belief. Although he rejected the dogmas of rationalism that we could gain certainty just through reasoning, he thus nevertheless lumped scepticism in with negative reactive beliefs, rather than recognizing its value in supporting provisional beliefs. Like the rationalists, then, he ended up using scepticism selectively – so-called 'mitigated' scepticism.

Hume was highly influential both on Kant and on later Anglo-American philosophers, who have uniformly continued with the same false assumption that scepticism is a threat to incrementally justified belief just because it systematically undermines absolute claims. From Putnam's discussions of brains in vats[19]

16   Ibid. ch. 3.
17   Descartes (1968).
18   Hume (1975).
19   Putnam (1982).

to Wittgenstein's assertion that 'certainty' could be claimed on conventional grounds,[20] philosophers have continued to quite unnecessarily see scepticism as a threat to which ingenious remedies need to be sought, even whilst selectively using its insights. Nor should the impact of these philosophical discussions on wider attitudes be underestimated. 'Scepticism' is almost universally treated as a selective negative position applied to specific issues (for instance, so-called 'Climate Change Scepticism') rather than a general ground of uncertainty, and thus lumped with *denialism* of the kind that selectively ignores the justicatory evidence for a position to which it is ideologically opposed. Only those who study philosophy tend to learn about general sceptical arguments, and then they tend to do so in a way that completely neutralizes their value, because they are framed as absolutizing irrelevancies.

In the process, then, a vital tool for provisionality, that could prompt further awareness of critical issues in even the most socially entrenched assumptions, has been sidelined and neutralized. Many opportunities for balancing feedback loops in thought, through the questioning of both positive and negative dogmas, have thus been lost, and the reinforcing feedback loops of dogma have instead continued. This phenomenon can also be associated with the limitation of judgement to the ideological level, where metaphysics is considered inevitable, even if some degree of universalizing takes place. One of the most striking indications of the lack of understanding of scepticism is found in the failure to understand or acknowledge the possibility of an interindividual level, as discussed above in 2.g, because of the confusion of practical justification with absolute belief that is characteristic of inflexible ideological thinking.

Overall, then, sceptical argument as a tool of balancing feedback has been almost universally perverted into a tool of reinforcing feedback. Entrenched beliefs continue to remain unquestioned, so their associated behaviours continue unchecked, because scepticism is used selectively to support them. Any challenges from sceptical questions are neutralized through the defence mechanisms that absolute belief has evolved.

20 Wittgenstein (1969).

3.g. Scientific Method and Scientism

> *Summary*
>
> Scientific method has developed from the early human practice of categorization and causal theory justified by experience. Through the ages this method has developed increasingly systematic observation and theorization, leading to a series of dramatic breakthroughs that have greatly helped humanity to address conditions. In modern times, however, over-specialization has made science subject to the same hijacking by reinforcing feedback as other cultural traditions, though somewhat less far advanced. This is shown in the slowdown and increasing distrust of science.

Scientific method is one of the crucial ways that human beings have adapted to the world, adjusting their beliefs in the light of new evidence, and thus creating balancing feedback loops. As human complexity has developed, we have also constantly improved the adequacy with which we learn from observation. However, the justification of beliefs from observation alone is no guarantee that learning and adaptation will take place, as confirmation bias means that we have a constant tendency to interpret our experiences only to fit our existing beliefs. In this case, then, science starts to create reinforcing feedback loops instead. Whenever the absolutization of our beliefs about the world prevents us from learning, we are then dealing, not with scientific method, but with dogmatic ideology – a dogmatic ideology that has often been called scientism. As we will see in this chapter, then, the history of science shows the same process of hijacking as other human cultural institutions like religion, socio-economic relationships, and sceptical philosophy, in which scientific method has turned into scientism. The hijacking is perhaps not yet as near-complete as it has been in religion and philosophy, but the signs that science may go the same way are worrying.

For the roots of scientific method, we should look back to the paleolithic, with the development of the language used to represent categories and causal relationships. In the observation of anthropologists, hunter-gatherers often have very detailed understanding of the animals, plants, and other phenomena in their environment, all of which is enmeshed with detailed technological understanding of their properties and uses. Here is an example from the Ryukyu Islands:

> Even a child can frequently identify the kind of tree from which a tiny wood fragment has come, and, furthermore, the sex of that tree, as defined by Kabiran notions of plant sex, by observing the appearance of its wood and bark, its smell, its hardness, and similar characteristics. Fish and shellfish by the dozen are known by individually distinctive terms, and their separate features and habits, as well as the sexual differences within each type, are well recognized.[1]

This is what Lévi-Strauss refers to as 'the science of the concrete'. He argues that in such science, we cannot entirely separate technological knowledge of usefulness from a wider and more disinterested understanding of our environment: 'we might well infer that animal and plant species are not known to the extent that they prove useful: they are found to be useful or interesting because, first of all, they are known.'[2]

Such earliest science is distinguished from modern science by its immediate relationship to the experience of everyone in the society – it is effectively indistinguishable from 'common sense', or from culturally shared understanding. Modern science, by contrast, is highly specialized and often incomprehensible to a large section of the population, shared and scrutinized only in seldom-read papers. In terms of the immediacy and application of understanding, then, a hunter-gatherer's 'science of the concrete' is far ahead of the average modern citizen's. Most modern citizens have nothing like such detailed engagement with tree bark or shellfish types. With that immediacy also comes a lack of the problems we are experiencing in contemporary discourse with trust in science. You are likely to trust what your grandfather told you about a given fungus being poisonous, because you have found him reliable in other matters and you recognize the full practical effects of ignoring his advice, but a modern doctor's advice about vaccination may fall on much less fertile ground. An immediate engagement with the objects concerned also makes it much easier to accept negative results of a kind that defy one's previous beliefs, because there has been no opportunity for 'knowledge' to be absolutized as such beyond its immediate practical application. In response to negative results one can then change one's theory. A Sumerian farmer called Shukallituda is recorded as having done this when his crops were destroyed by

---

1 Smith (1960), quoted by Lévi-Strauss (2021) p. 6.
2 Lévi-Strauss (2021) p. 11.

heat and wind. He prayed to the gods for guidance, and received the inspiration for a new strategy: to plant his crops in the shade.[3]

The development of science from these earliest beginnings was affected by the development of writing and by the emergence of some degree of specialization in the earliest civilizations – such as those of Mesopotamia, Egypt, China, and India. There were not yet any 'scientists' in these places, but there were priests, astrologers, and physicians who increasingly used systematic observation in pursuit of particular ends (such as healing or prediction), and who also developed the means of measurement of phenomena through mathematics, geometry, and the calendar. The fact that these results could be recorded meant that they were also passed on and shared more precisely and efficiently.

Already, then, here, we are getting into what we might call 'the science of the abstract' in contrast to Lévi-Strauss's 'science of the concrete'. Some empirical learning clearly occurred, but it was constantly interfered with by dogmatic assumptions. These were facilitated by the fact that science was being conducted by a limited and specialized group, beyond the scrutiny of others, and increasingly dependent on the written word, with its connotations of authority already discussed in 3.e. These early 'scientists' probably often employed an ideological level of judgement, but nevertheless one often structured by the projected archetypal beliefs that created 'religion' and 'magic'. For instance, the derivation of *predictive beliefs* rather than only *meaning* from astrology long interfered with the interpretation of astronomical observation. Scientism, and indeed pseudoscience, is thus not just a modern phenomenon, but was exhibited whenever ancient scientists remained unable to interpret their observations outside a dogmatic framework.

Perhaps the best-known ancient scientist was Aristotle, who was one of the first documented people, for instance, to make systematic observations of plants and animals. Nevertheless, his ability to use observation to challenge his prior framing is extremely patchy. Some of his cosmological beliefs, such as that anything beyond the level of the moon must be perfect, could be based on no observation, and persisted even beyond medieval times. Bertrand Russell famously pointed out that Aristotle erroneously believed that women had fewer teeth than men, but it did not occur to him to ask

---

3   Lawson (2004) p. 150.

his wife to open her mouth so he could count them.[4] The dawning of a new provisionality in ancient science, then, was nevertheless still accompanied by a strong influence of dogma that actually had the effect of turning people away from new understanding (especially after Aristotle himself became an overwhelming authority). It would be many centuries before Aristotle's emphasis on observation would be taken up again more systematically.

For the next advance in scientific method we need to leap to the early seventeenth century and the work of Galileo, who seems to have been the first person to use further recourse to empirical observation to confirm or falsify theoretical belief: particularly through the use of experiment. For example, he observed and then confirmed through experiment with different pendulums that the speed of oscillation of a pendulum is more or less identical regardless of its size.[5] His support for Copernicus's heliocentric theory of the solar system, which he supported from telescopic observations and defended against dogmatic *ad hoc* rejections,[6] is well known. In this Galileo was following through a new emphasis on the integration of the right hemisphere with the left to be found in the culture of the Renaissance, especially in Italy.

This approach alone, following the evidence where it leads, gradually filtered through Western civilization, and eventually led to extraordinary advances in the eighteenth and nineteenth centuries: for instance, Newtonian physics, the germ theory of disease, Darwin's theory of evolution, and the development of geology and astronomy so as to show the age of the earth. It's worth noting that all these advances occurred well before scientific specialization really got going. Darwin, for instance, was a gentleman amateur. It's also worth noting the degree of contemporary resistance they encountered, not just from 'religion', but more basically from scientism – that is, the absolutization of previous theories that were taken to explain the universe, regardless of the impact of new evidence. The conflict between Galileo and the Pope over heliocentrism, for instance, did not rest on any explicitly 'religious' source of information, but only on the fact that the Church had staked its authority on Aristotelian cosmology. In this sense it was operating

---

4   Russell (1952) p. 7, referring to Aristotle's *History of Animals* 509b.
5   Newton (2004).
6   I.4.d (pp. 110–11).

no differently from a 'scientific' establishment entrenched in a previous paradigm.

It's only in the twentieth century that modern science fully developed. Some of its developments, such as double-blind testing in medical research, do offer further boosts to methodological provisionality by offering further guards against confirmation bias. The development of quantum physics also obliged physicists to maintain a more provisional approach, as the scale at which they were working was one where limitations on the process of observation constantly interfered with the content of what was observed. Heisenberg's Uncertainty Principle, in 1927, formalized this recognition of the limitations of observation, in a recapitulation of sceptical argument. However, this has not prevented some thinkers from trying to appropriate quantum physics for dogmatic positions,[7] and in most respects the other branches of science (especially biology) remained untouched by it because they worked at a larger scale. As Iain McGilchrist comments, most biologists have continued to believe that biology is theoretically reducible to physics, whilst contradictorily ignoring the uncertainties that would imply.[8] An exception to this would be the development of systems biology by Maturana and Varela,[9] but this has remained marginal on the wider biological scene.

Since the early twentieth century, the *scale* of science has also expanded enormously, boosted by a massive rise in education, and hence of the probable number of people becoming capable of an ideological level of judgement. The number of scientists, and the numbers of scientific papers published, continue to rise rapidly.[10] However, despite (or perhaps because of) this, the rate of scientific progress seems to be slowing. Cowen and Southwood use a wide range of measures to evidence this slowdown: 'productivity growth, total factor productivity, GDP growth, patent measures, researcher productivity, crop yields, life expectancy, and Moore's Law'.[11] Another recent study, based on a very large survey of 45 million scientific papers, concludes that science is becoming less

7   See my response to a dogmatic philosopher of physics, Graham Smetham: Ellis (2011).
8   McGilchrist (2021) p. 431.
9   E.g. Maturana & Varela (1980).
10 Cowen & Southwood (2019) p. 22.
11 Ibid. p. 41.

disruptive, and the pace of scientific discovery is slowing down.[12] Given that substantial scientific discovery depends on effective engagement with conditions, and thus on the use of balancing feedback loops in judgement, the conditions that have contributed to this slowdown are not far to seek. All we have to do is look for the reinforcing feedback loops, and we can find these in the sociocultural context of science.

Many of the features of modern science are those of a social system: ever-increasing specialization, peer-reviewed journals, and an increasing trend for scientists to work in teams in laboratories. All of these developments are intended to enable more rigorous examination of evidence, by making specialized scientists increasingly aware of a very specific range of conditions, and relying on the critical awareness imparted by others within that specialism. However, this tends to ignore the likelihood of social biases in the specialized group, such as groupthink, ingroup bias, social proof, and false consensus.[13] Scientists are not immune from the tendencies to agree with the group (especially to advance their careers), dismiss the views of those perceived as outside the group, to use the group's view as a foundation regardless of contrary evidence, and to expect everyone else to have the same views as the group,[14] all of which are generally documented human tendencies. The more specialized and thus smaller the group, the stronger these tendencies are likely to become, creating a reinforcing feedback loop of expectations that are primarily social. Such pressures are also amplified by economic pressures such as competition for funding, and by the documented phenomenon of publication bias, in which (for both social and economic reasons) scientists are incentivized not to publish negative results.[15] The outcome of these pressures can easily be scientism: a continuing belief in the *results* of the social process of science rather than in the provisionality of its method. Results adopted in isolation from the continuing use of the method can easily become decontextualized, and thus dogmatic.

Specialism also creates fragmentation of the meanings[16] drawn on by scientists, as their jargon becomes incomprehensible to

12 Park, Leahy, & Funk (2023).
13 I.5.e; iv.3.e.
14 Allen & Howell (2020).
15 Dickersin (1990).
16 IV.4; iii.2.

non-specialists, and, most importantly, entrenchment of restrictive framing. Specialists are trained into certain assumptions, which then often go unquestioned for the rest of their careers, because people outside that specialism are not entitled to question those assumptions. This might have a more limited effect if scientists were also adequately trained as a matter of course in the philosophy of science, offering a wider critical context to scientific practice, but many are not. Some, indeed, actively despise and undermine philosophy, in effect seeing no acceptable grounds beyond the walls of their own discipline from which any criticism can be offered.[17]

As a result of this lack of contextual philosophical awareness, many scientists appear to have unquestioned metaphysical beliefs in materialism (or physicalism), determinism, the fact-value dichotomy, sometimes mathematical realism, and nearly always representationalism. Of course there are also a few scientists, unsurprisingly, who flip from this to the opposite metaphysical view, for instance by becoming theists. Even if such dogmatic beliefs are not formalized as such, or theoretically denied, they are often assumed in the absolutized format of 'knowledge' of 'nature' as the goal of science – terms that are still universally used (including in all scientific education) and that cannot be incrementalized, thus implying that the propositions of scientific theory are capable of representing reality.[18] 'Nature', which can have a valuable archetypal role in inspiring scientists,[19] is projected into a supposed representation of how things are. Such absolutized assumptions do not stand up to even-handed sceptical argument – but they are the result of the cultural entrenchment of selective scepticism in the scientific community.

If we compare the origins of science in concrete learning to its contemporary position, we thus see the onset of the same hijacking process we have seen in other aspects of human culture, but on the whole one that is less well advanced. Science is at a transitional stage in which many scientists are still working with a good deal of provisionality and integrity, despite being hindered by an increasing gathering of culturally entrenched absolutizations. For that reason, in my view, scientific evidence is still the most reliable

---

17   See, for instance, Lawrence Krauss's remarks: https://www.theatlantic.com/technology/archive/2012/04/has-physics-made-philosophy-and-religion-obsolete/256203/ (accessed 2022).

18   I.4.f.

19   See Ellis (2022) 6.a.

source of empirical information available, which is why I continue to cite it. Like any complex system, it will continue to work towards its main goals until it reaches a tipping point of absolutized corruption, when the reinforcing feedback loops will start to rapidly overtake the balancing ones. We already have increasing distrust, absolutized claims of scientific authority in response, and increasing political interference in science, but we have not yet reached the point where these tendencies overwhelm the long-established cultural trust of science – found for instance in the practice of medicine, in democratic political administration (amongst well-educated civil servants), and in education. Science is still one of the key areas in our culture where there is often an expectation of provisionality, but we should be very far from taking this for granted.

3.h. Technology: Shaped Things Shape Us

> *Summary*
>
> Technology has developed largely through a process of trial-and-error adjustment rather than of the application of formal scientific theory. Nevertheless, it can be hijacked by many reinforcing feedback processes: normalization, addiction, dependence on specialization and exploitation, and impact on the environment and on imagination. However, it is still possible to provisionally shape these things that have taken over and ended up shaping us, as in the use of renewable power technology.

In the earliest history of humankind, as mentioned at the beginning of the previous chapter, science and technology were indistinguishable from one another in a 'science of the concrete': that is, discovery both of causal properties, and of how to exploit them, through interaction with objects in the world. This could also be called 'trial and error'. Imagine, for instance, the creation of the first stone hand axe. Perhaps there was serendipity involved: someone picked up a rock that happened to have a fairly sharp edge and tried to use it for cutting. They were then struck by the idea of making the edge sharper, and tried various different other stones as sharpeners until they found one that worked. The same goes for finding the best stone to make further axes from, and developing the best strokes to work effectively with that material. At no point did anyone sit down to formally review or describe the properties of the rocks involved, or to design different possible models of axe and select the best one: rather there was a continuous interplay of embodied experience, increasingly consistent human motives, and the properties of materials that could be used to better fulfil human desires. In this way of working, because there is no abstraction of the design, it is almost impossible to absolutize that design. Rather, if the trial results in error, we just try something else – at least as long as we still have the resources to do so. The emphasis is on maximizing practicality as I discussed it in *Absolutization*, because the process is embodied, we take immediate responsibility for it, and we continue to work on improving its effectiveness.[1]

The divergence between science and technology, if there is one, lies in the abstract idea that science represents the universe in a

---

1   I.7.

merely 'descriptive' way, whilst technology harnesses this scientific description for human purposes. Like most such dichotomies, this covers up a much greater complexity. Not only does most technology *not* derive from prior science giving a formal description of conditions that technology then exploits, but it may also often be the case that it is science that plays catch-up by formalizing what technologists have discovered through practical trial and error. Nassim Nicholas Taleb gives a range of examples of how technology has been developed through tinkering: the jet engine, cybernetics, and the building of medieval cathedrals without the application of geometry.[2] Of course he also admits that there are some cases where technology has clearly been developed by applying pre-existing scientific theory, as in the case of the atomic bomb. However, it fits the complexity of the evidence much better to think of theories, where they exist at all, as providing us with further options, in the shape of meaningful possibilities or probabilities of how things will respond in practice, rather than complete representations of a state of affairs that is then merely 'applied'. To think of the science as entirely prior to the technology is a way of absolutizing the science – creating the scientism I discussed in the previous chapter, rather than applying scientific method.

Even so-called 'pure' science also has to be motivated, so that scientists spend time researching a particular issue and funders pay them to do so. Even if the scientists themselves have no particular application in mind as they pursue their 'pure' research, the funders may well do so. The 'purity' of science is thus obviously another absolutization. Instead of absolutely distinguishing pure from applied science, an application of the principle of incrementality can lead us to think of longer- or shorter-term applicability to human goals. Creating a representation of a state of affairs that may be useful in future is still a human goal, even if we can as yet have no specific idea of how that representation may be applied.

However, it is not only science that can start out provisional and be hijacked by absolute assumptions, but also technology itself. In the case of technology, it is not so much the way it is developed, but the reinforcing feedback loops created by its effects that lead us out of provisionality in our relationship to it. The history of human technology can show a variety of such effects, gathering pace as we go

---

2 Taleb (2012) pp. 221-3.

on. They can be associated with normalization, with addiction, with specialization, with exploitation, with the reshaping of our environment, and finally with the direct constraint of the human imagination. In all these cases, we started out by shaping materials for our own purposes and ended up with them shaping us back. That process is parallel to the move from provisionality in our judgements about technology to absolutization of its results, when it is no longer optional or negotiable but rather overwhelming.

The normalization of technology refers to the way that it fulfils our desires in a way that then rapidly becomes a normal expectation: part of the overall framing of our practical beliefs. If technologies are normalized, it makes it much harder for us to give them up, replace them, or amend them appropriately if they turn out to have unacceptably negative side-effects. This is closely related to a number of biases, including the *sunk costs fallacy,* due to which we're resistant to changing something that we've already put effort or money into (even when it ceases to work),[3] and *neomania,* which constantly focuses us on the next future technology to fulfil our desires, not on the ways that current technology is already fulfilling our desires.[4] It's also related to the *hedonic treadmill,* according to which new technology does not contribute to our pleasure or happiness in the long term.[5] We get short-term pleasure from acquiring a new gadget, but then rapidly incorporate it into our normalized baseline, so that the maintenance of the technology becomes necessary to maintaining our normal levels of happiness. We can see these effects readily in the launch of the latest smartphones, but there is no reason to think that they have not always applied to new technology in the past, even when the effects were much slower. Once we got hooked on ancient inventions like wheeled vehicles, stirrups or papyrus, for instance, it would have become difficult to imagine living without them. Although each new technology thus involves a balancing feedback loop, it then rapidly starts to feed a reinforcing one.

Normalization by itself could be seen as a sort of addiction, but this cycle can also be a more directly addictive one. Our chemical addictions to anything from caffeine and sugar to cocaine require the use of technological devices that have enabled people to process,

---

3   Kahneman (2011) pp. 343–5; V.5; iv.3.j.
4   Taleb (2012) pp. 322ff; V.5; iv.3.j.
5   Fujita & Diener (2005); V.6; iv.3.h.

transport, and consume these addictive substances. Technological devices themselves can also become not only normalized but addictive, as we constantly seek new dopamine hits from the experiences that the devices can give us. This is 'behavioural addiction' as opposed to chemical addiction, but the only difference here is that our bodies supply the addictive chemicals in response to repeated stimuli. Technological addiction is evidently rooted in all the other reinforcing feedback loops I have been considering throughout this book, but has only been noted recently – for instance as computer game addiction, internet addiction, social media addiction, or smartphone addiction.[6] It shows all the same basic features of addiction as those found in chemical addictions: for instance, as analysed by Mark Griffiths, these are salience (i.e. dominance in one's life), moments of euphoria, bodily tolerance, withdrawal symptoms, conflict, and relapse after attempts to end the addiction.[7]

The effects of specialization, already mentioned in relation to philosophy and science, and discussed in more detail in the next chapter, clearly apply to the development of technology as much as to these other areas. Whereas in ancient times, the resources and expertise for technological invention and application were widely diffused (requiring only a small amount of time, and perhaps a little wealth), the more complex the technology that has developed in modern times, the more it has required specialized resources and expertise. The ongoing normalized usage of technology has also produced demands for mass production, distribution, installation, advice, and maintenance that can only be fulfilled by specialists. So we now have aeronautical engineers, specialized washing machine repairers, computer software programmers, and second-hand car salespeople. This specialization has added a lot of fragility to our socio-economic system, because if one of these specialized elements fails it can rapidly block the operation of many other parts of the system. For instance, the failure of particular transport links due to technical failure can cause basic food shortages in areas that are dependent on that distribution chain.

The relationship between technology and exploitation depends initially on the concentration of resources that it demands. To create and apply a new technological device requires a concentration of

---

6 E.g. Ricci (2018).
7 Griffiths (1995).

labour and resources, whether this is an ancient plough or a new smartphone. Again, the more complex the technology, the more this requires socio-economic coordination. The makers of a new metal plough needed miners to dig out the iron ore, workers to refine it, and a smith to forge it, as well as farm workers to use it for the more efficient production of crops. The creation of all of the components of a new smartphone become much more complex than this, involving a long list of sources and processes for all the elements. Although the use of new technology for processing in the industrial revolution, and the development of mass production, made all these processes more efficient, the hedonic treadmill effect has tended to mean that this merely ramps up our expectations: the more new technology saves labour and resources, the more we require new labour and resources to produce further technology, either to patch weaknesses in the existing complex system, or to meet new, expanded, supposed needs. The role of advertising since the twentieth century has also helped to keep ramping up these new supposed needs.

It is this concentration of resources that tends to produce greater inequality through the development of technology, as Marxians have long noted. Only those with substantial resources already are able to apply the resources necessary to develop and apply new technology: hence the process of capitalistic investment and profit, which makes the rich richer and the powerful more powerful. On the other hand, the workers who contribute to the process are rewarded only in proportion to the scarcity value of their skills (and this assuming that the labour market is sufficiently regulated so that reliable information about different workers' skills can be easily compared). In ancient times, these workers might have been slaves – though some slaves had higher status than others. In modern times, they are usually employees, whose reward is subject to the operation of the market. Technology, then, may overwhelmingly reward investors, skilled workers, and consumers whilst nevertheless increasing the exploitation of less skilled workers, even when the market is operating in a relatively 'free' fashion. Whilst technology has theoretically held out the promise of freeing all humans from drudgery, in practice it rarely succeeds in this. In practice the reinforcing feedback loops of exploitation can be added to those already mentioned. These reinforcing feedback loops of exploitation constantly interact with the socio-political ones already discussed in 3.d.

The effects of technological change on the environment are widely understood, so only need brief summarizing here. New technologies are likely to use new resources (raw materials, energy, and perhaps land) in their creation and in their operation, and they may also produce pollution or other environmental side-effects. All of these introduce reinforcing feedback loops that may be accelerated by the spread of the technology and by increasing human dependence on it. The earliest human technologies, such as farming, disrupted ecosystems to some extent (for instance through deforestation), but were also relatively sustainable because less intensive. However, the acceleration of intensive farming has multiplied all these effects, to the extent that half of all habitable land (29% of all land) is now used for agriculture, with an accompanying massive impact of reduced biodiversity.[8] Newer technologies, such as computers for instance, depend heavily on mining, manufacturing, and electricity generation, as well as the technical input of designers and programmers. All of these processes are depleting finite resources, stressing renewable resources (such as water) beyond the point where they can renew themselves, and disrupting ecosystems on which human and animal welfare is reliant, whilst their output of greenhouse gases threatens to make the earth uninhabitable through accelerating global warming. A simple measure of the unsustainability of these self-reinforcing processes is provided by Earth Overshoot Day, which measures how much our use of resources in the world as a whole exceeds its capacity, by identifying a day each year when we have used up the resources available for that year: in 2022 we had overshot by 28 July.[9]

The reinforcing feedback loop of human technological development is modelled in economics by growth, which broadly speaking is a measure of the processing of resources from the wider environment into human wealth. Our financial system largely depends on growth as a normal state of affairs, so that the cost of investment in new technology and its diffusion is recouped by the investors. Without growth we are burdened with public and private debt – another reinforcing feedback loop that can overtake governments, organizations, and individuals. It's already been clear for some time that the economic growth model on which we rely is unsustainable,

---

8 https://ourworldindata.org/land-use#how-the-world-s-land-is-used-total-area-sizes-by-type-of-use-cover (accessed 2022).

9 https://www.overshootday.org/ (accessed 2022).

and that it's very likely to reach a tipping point and to crash, in co-dependence on the loops of normalization, addiction, specialization, and exploitation that I have already noted above. In this, it simply follows the pattern of any complex system that has become stretched beyond its capacity to self-adjust.

Just as the development of technology in ancient times began more as a provisional response to the environment, however, there is also no reason why it cannot be used to introduce balancing feedback loops that will correct the accelerating cycles of self-destruction. This is of course the thinking behind the use of technology to create renewable energy (wind turbines, solar panels, etc.), to recycle resources, and to create acceptable sources of protein from micro-organisms in vats using a tiny fraction of the huge ecological resources used to rear animals for meat. It is unlikely that such adaptive technology can ever lead humans into a completely homeostatic state in which they have no long-term conflicts at all with their environment (the production of solar panels and wind turbines still requires mining and manufacturing). Nevertheless, it can introduce balancing feedback loops that can hugely slow down the reinforcing ones, and move the point of ecological self-destruction much further into the future. The pattern of turning reinforcing feedback loops into balancing patterns of adjustment is no different from the one used by all forms of life, as tracked through the first section of this book. Life itself is a reinforcing feedback loop (see 1.a) that constantly uses resources and always threatens to spiral into unsustainability, but it can also always modify itself to new conditions by whatever forms of complex adjustment are available to it. We should thus avoid the nirvana fallacy (rejection on grounds of imperfection) in response to green technology, just as we should avoid it when faced with any life-form (a rabbit, say) that might potentially multiply itself into oblivion in the wrong systemic context.

The final impact of technology is the way in which it can have the effect of restricting options by impacting the imagination. Any technology could conceivably have this effect to some degree, by constricting our capacity to consider actions without it. We are far less likely to seriously consider walking 100 miles to a place we want to get to, for instance, if we could ride a chariot, a train, or a car instead. In the case of technologies that apply energy to make movement or other activity easier, though, we could argue that this effect is offset by the new options we start to consider because of the technology: we could not travel to Bali without aeroplanes, and we would not have considered making AI art without AI art-producing software. These new options have their costs, but they also can also add to and integrate meaning in our lives.[10]

Where the restriction of our options becomes much more threatening, though, is in the realm specifically of communications technology. I have already discussed the ways that the development of writing interacted with idolatry (3.e). The next major historical development of communications technology is probably printing, which supported the Protestant Reformation by allowing the release of a surge of popular printed broadsheets and pamphlets, as well as accessibility for the Bible.[11] In this, then, popularized religious beliefs started to replace ones imposed by the Catholic Church hierarchy, and offering a new kind of reinforcing feedback loop for the social spread of beliefs. Similar effects then accelerated with the rise of broadcast media – radio, television, the internet, and particularly social media and smartphones.

That such media could be used effectively for propaganda purposes was already well-established in the early twentieth century. Radio, for instance, was used effectively in early twentieth-century Germany to spread support for Nazism through antisemitism. Such propaganda messages merely needed to speed up the group-binding effect of absolutization,[12] providing a dominant message that excluded alternatives to a large number of people simultaneously. It has been found, however, that Nazi antisemitism on the radio was more readily accepted in areas that already had greater antisemitic tendencies.[13] Once accepted, then, the absolutizing message had a tendency to exclude alternatives and take over, but it could only take root in the absence of sufficient alternatives. We see this same tendency continuing today in the spread of conspiracy theories through social media: the medium merely provides rapid simultaneous transmission that speeds up the spread of the absolutizing beliefs, but once sought out and accepted they rapidly become dominant through all the mechanisms of absolutization. This is particularly apparent in the polarized effects that accompany

10 IV.5.
11 Rubin (2012).
12 I.5.e.
13 Adena et al. (2015).

conspiracy theories[14] (fitting the option exclusion and dichotomization of absolutization[15]), and in the 'echo chamber effect' by which social media emphasizes reinforcing feedback loops of group association judgement.

Once again, there is no reason why these particular forms of communication technology could not be used provisionally. However, the effect of their use given the human capacity for absolutization is often one that produces further reinforcing feedback loops in interdependence with all the other ones we have considered. Technology on the whole not only produces accelerating reinforcing feedback loops, but even forces our reflective thinking about it into further reinforcing feedback loops. This process, however, began from a provisional basis to which it can return.

---

14 Cinelli et al. (2022).
15 I.1.d.

3.i. Specialization and Over-specialization

> *Summary*
>
> Specialization is most basically an adaptive biological process that unavoidably reduces options. However, it starts to create reinforcing feedback when we absolutize our specialized niche. The most basic human specialization is that of gender, which started provisionally but became absolutized in patriarchy. Class over-specialization has developed new forms under capitalism, creating the necessity for 'recreation', while professional over-specialization constricts thinking. Most balancing responses to this have been superficial in practice.

Specialization is the process involved whenever we move from a larger range to a more specific focus of attention over the longer term. That focus of attention then shapes our actions, our environment, and us. It enables us to address more limited conditions better in the short term, but comes at the price of limiting our ability to address wider, longer-term conditions.

Before the evolution of human beings, specialization can be seen as a general biological process: a plant or animal that focuses on particular ways of meeting its needs in particular circumstances then becomes better adapted to those conditions. This can apply both at an individual level (for instance, the stretching of branches in one direction, or the development of muscles through usage) and at a species level (genetic and epigenetic inheritance of specialized characteristics). Every time an organism becomes specialized to a particular niche, however, it sacrifices its adaptability to some extent, as it will lose the capability to meet the conditions in other possible niches. The example of the giant panda here is notorious: by specializing only in certain types of bamboo, it has placed itself at the mercy of changes in bamboo growth levels, and thus greatly restricted its ability to adapt to new circumstances, including those of global warming.[1] One of the greatest strengths of humans (like rats) is that we are generalists, readily able to exploit new opportunities in very varied environments. The tendency towards specialization within human societies might be seen as a compensation for that generalism as a species.

---

1   Renqiang Li et al. (2014).

Specialization by itself is an adaptation to an environment, requiring provisionality. However, once a process of specialization is completed and we are settled in a niche of conditions, it can readily become a source of absolutization, limiting our ability to engage with new conditions, because our beliefs are fixed to a limited range within the specialized niche. This is the point where specialization becomes over-specialization. We cannot entirely refuse to specialize, any more than we can refuse to live or refuse to grow up: all of these processes gradually limit our options, and involve reinforcing feedback loops continuing to operate in the background. Specialization, to some extent, is merely the exercise of optionality. However, the point of judgement lies in how much we allow that specialization to restrict our future options: whether we see the specialization as the whole story, or only as part of a wider context.

The most basic human specialization, rooted in the biological distinction of sex, is that of gender. Such specialization may have emerged in the earliest human history simply as a result of the continuing exercise of different strengths – for instance, men, having generally greater physical strength and aggression, would tend on average to do tasks that required it, such as hunting and defence against invaders. Women, being particularly equipped with both breasts and caring instincts, would tend to specialize in childcare. The more each performed these differing roles, the more they were likely to diverge, even though men were still capable of looking after children, and women were still capable of hunting and defending themselves. In this sense, basic specialization within human societies reflects that between the left and right hemispheres, discussed above in 1.g. As with the hemispheres, specialization forms us and limits our options, but is not destiny. As with the hemispheres, there are also many tasks that might equally be performed by either gender, or where the gender allocation has varied greatly between cultures. There have also probably been people who defied the binary division between genders, in one respect or another, throughout history.

Hunter-gatherer societies have often had strong gender specialization, even when they had little specialization in other respects. As Kuhn and Stiner write:

> *The basic generalization from which we begin is the nearly axiomatic division of subsistence labor by gender and age documented in virtually all recent foraging*

peoples: put most simply, men hunt, women and children gather. More precisely, males are normally responsible for obtaining large terrestrial and aquatic animals, whereas females and, in some cases, juveniles focus their efforts on vegetable foods and smaller animals in addition to helping process the big animals. This basic form of gendered division of labor is expressed both in ideology and in practice in nearly every ethnographically documented foraging group.[2]

They go on to qualify this only by pointing out that, while general, it is not absolute at the individual level:

Individual departures from this pattern are numerous, according to preference, context, and individual circumstances. On a given day, almost anyone in any foraging society might take advantage of opportunities to bag a big animal or to pick berries, albeit within the limits of their physical capacities. Among high latitude hunter-gatherers, widowed women or daughters in families without sons could become successful and habitual hunters.... Likewise, male hunters often (though not always) take vegetable foods or small game when a good opportunity presents itself.[3]

This complexity can be reinforced by other findings that cut across the previous assumed rigidity of specialization: for example, eleven burial sites from the late paleolithic have been found across the Americas in which hunting weapons were buried with female bodies.[4]

Like the development of farming or other technology in early human history, then, it seems fair to suggest that gender specialization was a provisional affair, taking advantage of relative strengths but not limiting the options for individuals' roles very much, within the bounds of practicality. Nor should it be assumed that in hunter-gatherer societies, this specialization is necessarily the basis of an absolutizing patriarchy. As Graeber and Wengrow comment, many early societies have such distinctive and separate domains for men and women that they could be seen as patriarchal in one sense, but matriarchal in another. Men may tend to dominate political offices, but this does not mean that they always wield all the power, whereas women are much more likely to control land and its produce.[5]

The development of gender specialization into patriarchy may have accompanied the rise to power of absolute male rulers in early urban civilizations. This involved not just the potential for male

2 Kuhn & Stiner (2006).
3 Ibid.
4 Cirotteau, Kerner, & Pincas (2021) pp. 134–6.
5 Graeber & Wengrow (2021) p. 74 and note 63.

dominance through violence, which of course has always been present, but the normalization of it. Graeber and Wengrow argue that it may well have been due to the confusion of the roles of care and domination that occurred in early urban civilizations when widows and orphans came under the paternalistic care of kings. At that point, the development of male-dominated administrative organizations (such as temples in ancient Sumeria), started to take oversight of the previously female preserve of care and nurturing.[6] Socio-political discourse from this point starts to adopt what George Lakoff calls the 'stern father model'[7] rather than only the nurturing mother metaphor, with the king as the stern father of the nation. From this model, it becomes culturally acceptable for the individual man to also become the king of the household. From this point, then, we could see gender specialization as losing a crucial element of provisionality and becoming at least partially a form of oppression, to be rectified only with the rise of feminism in modern times.

This loss of provisionality in gender specialization is of course also accompanied by the rise of other kinds of specialization in ancient urban civilizations. At this stage these are not so importantly specializations of profession (priest, scribe, administrator) as of class. The separation of those who have power and wealth from those who are merely producers is the key one, with the latter often becoming a class of slaves or serfs in ancient and medieval societies.

The Indian *varna* or class system separates out three classes with some power and wealth from producer and enslaved classes – these being priests (*brahmins*), warriors (*kshatriyas*), and traders (*vaishyas*) as distinct from the fourth class of *shudras*.[8] This, then, is primarily the specialization of gross inequality mentioned in 3.d above, in which it is overwhelmingly the oppressed class that loses options and is moulded to limited tasks. Other civilizations have had varying levels of mobility between the different professional specializations of those with higher status, but the Hindu tradition absolutizes even this type of specialization in the form of caste duty (*dharma*). In the *Bhagavad Gita*, often seen as the most revered text of Hinduism, the divine figure of Krishna preaches at length to the

---

6   Ibid. pp. 517–21.
7   Lakoff (2002) ch. 5.
8   Sanctified in the myth of the dismemberment of the cosmic man in the *Rig Veda* (10.90.1–16), and further rigidified in the *Law Book of Manu*.

kshatriya Arjuna that he should fulfil his caste duty to fight, even when this means killing his relatives.[9]

We can thus see the argument between Brahminism and the alternative non-Brahminical traditions in India (such as Buddhism and Jainism) as partly one about the rigidity of specialization. Whilst the Hindu caste system developed increasingly rigid forms overseen by Brahminical power, the Buddha challenged this by arguing that *dharma* was universal morality, not dependent on a specialized class or role. In this, then, he was applying the Middle Way by urging a return to provisionality from specializations that had become repressive ends in themselves. Similarly, in Europe, one of the key developing points for greater provisionality during the Renaissance was the improvement of social mobility, which had come with the development of towns and artisanal guilds. In Tudor England, Thomas Wolsey, the son of a butcher who rose to become cardinal and chancellor, and Thomas Cromwell, the son of a blacksmith who eventually also became chancellor, strongly illustrate this trend.

However, the gradual loosening of oppressive social specializations from the European Renaissance was followed by a renewed tightening of them during the industrial revolution. Whilst peasants became free from serfdom and could more easily move to the cities in search of a better standard of living, the 'free' exchange of labour for pay rapidly became a new form of oppression in the developing factories of the nineteenth century. As previously mentioned in 3.d, the mere idea of 'freedom' in one respect here becomes an absolutized substitution for the experience of it, with workers forced by the restriction of options in the economic system to accept very low wages in return for toiling in the grip of the pitiless rhythms of a machine. During the twentieth century this oppressive structure was then increasingly moved to workers in developing countries, even whilst the welfare state improved the state of inequality within developed countries.

The impact of industrial capitalism not only accelerated the specializing distinctions between classes, but within the lives of individuals also created a specialization of time allocation. Whilst oppressed people of earlier ages simply had lives of non-stop toil, the industrial worker began to operate in timed circumstances. The development and spread of clocks meant that expectations of

---

9   Zaehner (1969); also see Ellis (2022) pp. 205–6.

time became far more precise, with workers obliged to spend set amounts of time working for their employer, also leaving time aside for 'leisure'.[10] The distinction between work and leisure has the effect of specializing time periods within our lives for repressive winding up and compensatory winding down: reinforcing feedback loops of tedious, alienating activity followed by balancing feedback loops of 'recreation'. Recreation is, of course, not necessary if we are wholly engaged in what we are doing, rather than repressing the part of ourselves that doesn't wish to be doing it until it can come out later. What we are doing is then not 'work' in the necessarily alienating sense that we have often come to associate with it, and in that context becomes balancing feedback – as I will discuss further in 4.k. There does not need to be any dichotomy between work and leisure, but it is specialization that creates that dichotomy by repetitively bending our bodies to one limited kind of task as 'work'. In a world shaped by that dichotomy, of course we need recreation, but it would be better still to be free of it.

This history of the ways that specialization has become oppressive forms a necessary background to the forms of over-specialization I have already complained about in 3.f, g, and h – the specializations of philosophy, science, and technology which have had such constricting effects on human thought. These may seem like merely intellectual specializations, still leaving specialized practitioners with a lot more options compared to factory workers, and to a large extent this is correct. The most hidebound scientific academic, for instance, could still probably enjoy a more comfortable and reflective life than the average factory worker. However, there are also advantages for those who have to exert bodily effort in their work, even if in a relatively repetitive way, that are not open to those who spend a lot of their days sitting at desks. The specializations of intellectuals are creative of more reinforcing feedback loops both because of their tendencies to encourage disembodiment, and because of their wide long-term influence over the assumptions of their society.

In responding to specialization, we need to beware of merely formalistic antidotes to it. These might include 'equal opportunities' that do not address many of the conditions that maintain absolutizing assumptions in patriarchy, or many of the conditions that

---

10   Martineau (2015).

hold working-class people back from individual development. In education in England, a typical response to the recognition of over-specialization can be illustrated by the strategy of the authorities of including 'General Studies' to broaden the over-specialized A-level curriculum for 16–19-year-olds. Of course, this additional study made almost no difference to the students' future prospects, so was not taken seriously.[11] At the intellectual level, formalistic antidotes may include 'interdisciplinary' work that actually fails to challenge the restrictive assumptions of some of the clusters of disciplines (such as social sciences) that are likely to inform it. In such cases, the constricting assumptions behind specialization maintain their socio-economic predominance, so continue even when some of the more superficial barriers between specializations are removed. We thus need to address specialization at the philosophical and psychological as well as the socio-political level.

It was Carl Jung who offered one of the strongest and deepest challenges to over-specialization and its effects during the twentieth century, in the insights that led to his recognition of the *compensatory* anima and animus archetypes. Although these later tended to focus on the effects of gender specialization to the exclusion of other kinds, Jung's earliest work in the Red Book shows a recognition of a more general specialization problem.[12]

> Without balance you transgress your limits without noticing what has happened to you. You achieve balance, however, only if you nurture your opposite.[13]

The 'opposite' consists here of the repressed desires and beliefs that specialization of all kinds leaves us with – which we encounter through the 'soul'. Jung mentions not only the femininity of men and the masculinity of women, but also such examples as the simple-mindedness of the clever and the authority of the servile.[14] It is only due to over-specialized repression that we need compensatory practices, sometimes including psychotherapy, to address it.

---

11 https://www.theguardian.com/education/2015/jul/14/is-general-studies-a-waste-of-time (accessed 2022).
12 Ellis (2020) 7.a.
13 Jung (2009) Liber Secundus 8, translation p. 263.
14 Ibid., Liber Primus iii(r), translation pp. 235 and 237.

3.j. Administration and Bureaucracy

> *Summary*
>
> There is historical evidence that, contrary to long-held assumptions, provisional administration is possible without the absolute imposition of top-down authority and rule-following rigidity. Some ancient models separated administration from sovereignty, as also proposed by anarchists – an approach that requires suitable psychological conditions. Modern bureaucracy, though, often follows its own self-justifying left-hemisphere 'rationality', whilst political reactions against it have failed to reframe, themselves relying on bureaucratic managerialism.

'Administration' is normally a neutrally-toned word, whereas 'bureaucracy' has become pejorative. We can readily recognize that 'administration' is necessary to make sure that any organization does its work with a reasonable degree of consistency. Funds and resources have to be made available, those working for the organization need some degree of supervision to ensure they are doing their work appropriately, however much they may be trusted, and records have to be kept to ensure that everyone is treated fairly and efficiently when they interact with the organization. 'Bureaucracy', however, is administration that lacks critical awareness of itself, and has started to multiply in endless reinforcing feedback loops: to get the necessary resources, people now have to fill in fifteen forms that are countersigned by ten different people; helpful supervision becomes micro-management with burdensome recording requirements that now greatly reduce the effectiveness of the work. Bureaucracy is one of the curses of modern times.

To track how the former has too often turned into the latter, we need to start with the origins of administration. Until recently, historians seem to have assumed that administration necessarily accompanied the rise of the state. Perhaps this assumption went with the unquestioned belief that bureaucracy was the special vice of states (a belief often associated with the anti-state libertarian right in recent times), but if we can separate administration from states, it may be easier to recognize how much both administration and bureaucracy are the functions of any large and organized group of people, including businesses, charities, and even groups working in highly cooperative and egalitarian ways without strong central authorities. If we draw the sting from administration by showing how it is not necessarily always associated with power, but can be used in provisional and contextualized ways, then we can see that there is optionality here too. A helpful human function has, once again, been hijacked for absolutizing ends.

Graeber and Wengrow are highly illuminating on the ancient origins of administration. Instead of assuming that any absolute ruler of a territory (that is, a king) must necessarily also administer it, they have identified many examples of rule without administration. When a king rules, he has a monopoly on the use or threat of violence to maintain control, but in order to effectively extend that control over the activities of a population over an area of land, he must also access and control information about their property and activities, as well as of the legal framework he wishes to apply to them. The channelling of that information requires administrators – civil servants or similar functionaries who do not just specialize in violent enforcement, but rather in understanding and responding to the conditions. The early rule of kings without administration can be illustrated by the rule of the king (or 'sun') of the Natchez, a Native American group in Louisiana. The Natchez king evidently had total power within his immediate environment, including being able to order arbitrary executions and have them carried out. However, beyond his immediate physical presence, the Natchez would simply laugh at and ignore any order from the king that they did not like.[1] It took the more consistent development of civilizations such as Ancient Egypt, to ensure that the absolute ruler's wishes were indeed carried out throughout his realms, due to the presence of administrators who ensured consistency. Graeber and Wengrow suggest that this extension of power was based on the extension of kinship rituals to encompass the whole people of the territory.[2]

On the other hand, there is also evidence of administration without sovereignty in early societies, with systems being set up to store, share, and convey information for what seem to be mutual purposes, or at least ones where leadership is provisional and conditional. Neolithic settlements of considerable size (around 1000 houses) have been excavated in Ukraine and Moldova, dating from

---

1 Graeber & Wengrow (2021) pp. 156-7 & 368.
2 Ibid. p. 402.

around 4100 to 3300 BCE, in which the people seem to have administered themselves without centralized control.

> There is no firm consensus among archaeologists about what sort of social arrangements all this required, but most would agree the logistical challenges were daunting. A surplus was definitely produced, and with it ample potential for some to seize control of the stocks and supplies, to lord it over others or battle for the spoils, but over eight centuries we find little evidence of warfare or the rise of social elites.[3]

The same applies to a wide range of societies in the early neolithic, including villages around the Near and Middle East, the early Sumerian cities, and the Indus Valley civilization. Very often, archaeologists seem to have simply expected there to be a king because there was an urban settlement requiring complex administration, so have read this into the evidence: for example, a cloaked figure found in the Indus city of Mohenjodaro has long been interpreted as a 'priest-king', when there is no particular evidence of kingship at all.[4] The confirmation bias has assumed that centralized authority accompanied administration as a default, without any demand for evidence of it. Amongst some contemporary groups, too, there is positive evidence of administration without sovereignty:

> Today, it is quite commonplace – for instance, in parts of Africa and Papua New Guinea – to find initiation ceremonies that are so complex as to require bureaucratic management, where initiates are gradually introduced to higher and higher levels of arcane knowledge, in societies that otherwise have no formal ranks of any sort.[5]

According to the anarchist thinker Peter Kropotkin, writing in 1902, such non-absolutized social organization not only occurred in these ancient times (what Marx and Engels referred to as 'primitive communism'), but also continued well into the Middle Ages and beyond.[6] For instance, he regarded the medieval guild system as one based on *mutual aid* rather than centralized authority, and argued that it was the rise of increasingly centralized bureaucratic states in the sixteenth century that ended much of this mutual aid system. Nevertheless, he continued to find it in much European village life, and more recently in workers' unions. The difficulty in

---

3   Ibid. pp. 293–4.
4   Ibid. p. 318.
5   Ibid. p. 365.
6   Kropotkin (1989).

all such anarchist writing seems to lie in the assumption that non-absolutizing forms of social organization depend only on socio-political arrangements, rather than on psychological conditions. Often, too, there is a dogmatic belief in the innate goodness of human nature. However, as we saw in 1.i, there is a biological basis for provisional relations of solidarity between people, as between other animals, in right-hemisphere coordination. Mutual aid is thus a matter of degree in all organizations throughout history, depending on how much people's relationships were dependent on entrenched left- or right-hemisphere dominant ways of relating. Nevertheless, there are socio-political arrangements that can better help us to create the conditions for genuine solidarity and mutual aid, by avoiding the entrenchment of absolutization in the judgements that create social relationships. This will be part of a more detailed discussion of political philosophy in volume IX of this series.

For the moment, however, it is worth mentioning the features that Kropotkin attributed to genuinely anarchist organization in contrast to those of top-down rule, to get an idea of the common cultural features of administration without sovereignty.[7] Kropotkin thought that hierarchical authority leads to a lack of autonomous freedom for individuals. Although Kropotkin admitted that anarchism could not make people perfectly free because of the need to cooperate, a non-repressive administration would maximize individual autonomy. One could add in here the beneficial psychological effects of such autonomy in enabling individual integration. He also thought that hierarchical authority led to corruption of a kind that could be avoided through mutual aid: this can be related to the loss of responsibility if we have a less integrated basis for cooperating. Thirdly, he thought that hierarchical authority was ineffective, in contrast to the ways that full individual responsibility of people working together in harmony can operate: I have offered a detailed argument supporting this in *Absolutization*, in relation to the ways that absolutization interferes with practical effectiveness.[8] This also relates to the ways that hierarchical authority leads to conformity and obedience, repressing initiative and creativity of a kind that he found commonly available amongst people who had not been

7   De Geus (2014).
8   I.7.d.

ground down by repressive authority. Instead, then, he thought that society could be administered through freely associating and self-regulating units. Individuals freely associate in smaller groups, and the smaller groups freely associate with other such groups to form a federal network. The basis of joint action would then be contractual agreement through a decision-making tree.

There are difficulties with the kind of administrative systems that Kropotkin envisaged when put into practice, however, as has become evident in various 'flat' organizational systems that have been proposed and trialled since Kropotkin's time. These include the ways that it can interfere with leadership of the kind that is needed to develop new approaches and strategies in practice. Highly decentralized systems can become rather conservative, and unable to compete with those that are more strongly led from the front. The decentralized structure can also greatly add to the weight of administration rather than reduce it, because of the need for everything to be negotiated and recorded transparently. One of the most recent attempts to develop an effectively balanced administrative structure is Brian Robertson's holacracy.[9] A holacracy is organized into 'circles', which are self-governing and work out their own ways of achieving agreed goals, but these goals are set by the wider organization, or circle of circles. Individuals have agreed roles within the circles that are kept under regular review. Decisions are made in both wider and more specific circles through 'integrative decision-making', which is not merely majority voting or consensus, but a process of trying to match information to the overall needs of the organization. Holacracy avoids the conservatism of more traditional forms of decentralized governance, but is focused more on the needs of 'agile' businesses or other organizations in a competitive market than on providing a general model for social administration.

The question of what social form non-absolutized administration should take, then, is an ongoing one that has evidently continued from ancient times into the present. However, the fact that such debate and experimentation is still taking place offers general evidence that provisional administration in support of mutual aid is possible. It is not inevitable that we need to adopt absolutizing shortcuts in group organization as soon as we reach a magic

---

9   Robertson (2015).

'Dunbar number' of around 150, beyond which we can no longer practicably maintain relationships of mutual trust without 'cognitive overload,'[10] because there are ways of delegating and thus extending that trust. Clearly, the exact form that provisional forms of administration take itself needs to be provisional. An application of the Middle Way, in constantly balancing between the demands of addressing wider conditions on the one hand and of maintaining individual and social integration on the other, is clearly required.

Once we have established that wider point of the emerging possibility of non-absolutizing administration, though, we are left with the historical increase of bureaucracy. There are several different senses in which the term 'bureaucracy' is used,[11] either neutrally or pejoratively, but here I am using it in contrast to 'administration', to mean administration in which absolutization is culturally entrenched and reinforced. This kind of administration is 'monocratic', as Max Weber called it in his influential discussion of bureaucracy dating from 1947, meaning that it appeals for its legitimacy to a set of unquestioned legal norms that provide a unifying top-down authority.[12] Such bureaucracies work on the basis of a consistent set of absolute rules, a set of prescribed roles for the staff who are separated from their individual identities, and a dependence on written records that provide the basis on which the rules are applied. They are also geographically and culturally centralized, without adaptation to the variations within the larger or smaller area they cover. Bureaucracy in this sense has hijacked administration, just as all the other absolutized cultural forms have hijacked provisional ones, as seen throughout this section. However, reactions against bureaucracy that fail to challenge its absolutizing assumptions also tend to unwittingly merely reproduce them, in a pattern that should already be familiar to any student of the Middle Way.

Max Weber offered an apparently positive assessment of this type of bureaucracy as an application of 'rationality' (although it has also been suggested that he has been misinterpreted[13]). It does indeed exhibit 'rationality' in the sense of a consistent application of certain shared assumptions, and in this sense being highly dependent on the left hemisphere. As long as the assumptions being applied

10   Dunbar (1992).
11   Farazmand (2009) p. 5.
12   Weber (1947) pp. 324–41.
13   Gajduschek (2003).

are adequate to the conditions that are operating, then, such consistent 'rationality' will be helpful. For example, people will remain confident that the bureaucratic system will fulfil its functions fairly, whether it is assessing taxes, dispensing benefits, enforcing planning regulations, or (in a private company) assessing a new candidate for employment. However, this consistency is applied at the expense of rigidity. There will be no adaptation to the needs of the individual citizen, or the individual bureaucratic employee (possibly creating further conflicts within these individuals). There will be no allowance for the initiative of the citizen who has new ideas and practices that are actually for the public benefit, if they do not accord with the rules. Following the left hemisphere's tendencies, it will consistently apply the policies that fit its version of 'reality', in the utter conviction that this version is 'true' and 'right', but without the capacity to adapt its beliefs, because any capacity for adaptation within the bureaucracy conflicts with its very mode of operation.

Instead, adaptation is left to the political level above the administration, which may or may not adapt in an appropriate and timely way, but whose decisions will nevertheless be inflexibly applied. Whether political decision-making is provisional is, of course, another discussion (to be pursued in volume IX). However, even if we assume that political decision-making is always provisional, this decision-making may be constantly undermined if it is applied by an absolutizing bureaucratic system in which state employees have no discretion to adapt their approach to the context. This may become particularly obvious where individual motives are in harmony with the spirit of a democratically-enacted law, but the letter of the rules being applied actually prevents those individuals acting on their motives. One example of this may lie in planning laws that are intended to protect the environment, for instance by insisting that agricultural land is not developed. However, someone then wanting to build a house that enables them to work on their land in a more environmentally-helpful way may then be prevented from doing so by such regulations. At the same time, large-scale developers may be able to evade the same level of scrutiny through corruption, perhaps cultivating the support of politicians, or employing lawyers to follow the letter of the law whilst disobeying its spirit.

Such bureaucratic rigidity dates back to the systems of civil service established in ancient empires, particularly the ancient Chinese and Persian Empires. In the ancient context, however, there do also seem to have been some attempts to balance the absolute bureaucratic control of the system with some adaptability through the virtues of the bureaucrat. That, at least, would be one way of interpreting the institution of the Chinese imperial examinations that controlled entry to the civil service on the basis of knowledge of the Chinese classics.[14] On the one hand, this has seemed to many Western observers to be an absurd irrelevance, but on the other, it can be seen as an attempt to introduce a wider integration into the character of those with administrative responsibilities. This integration of character is indicated in the Confucian ideals of how all administrative relationships should be conducted, in accordance with Confucian virtues of benevolence, righteousness, propriety, wisdom, and trustworthiness.[15] The personal character of civil servants or other administrators everywhere can, of course, allay the rigidity of the system, but only insofar as the system itself allows them any degree of personal discretion.

In the modern period, however, bureaucracy developed in pace with the rise and development of the state, which meant both the consolidation of state power within its borders and the extension of state responsibilities. With the rise of capitalism, other kinds of organizations also developed with large bureaucracies, such as private companies: for example, the East India Company, which effectively ruled India from 1757 to 1858. Nevertheless, it was state bureaucracy that multiplied most obviously.

It was C.N. Parkinson's writings on bureaucracy that began to make it clearer how this bureaucratic expansion could be following a reinforcing feedback loop, not merely due to social, economic, or political development but also to a momentum of its own. Parkinson noted, for instance, that the numbers of administrators in the British Admiralty rose by 78% between 1914 and 1928, at the same time that the number of ships had declined by 67% and the number of men in the navy by 31%. Parkinson accounted for this with a general law that 'Work expands so as to fill the time available for its completion'.[16] This observation may not always be correct at an individual level, where it assumes an innate human tendency to procrastination that may often, but not always, apply. However, in the case of administrators in an organization, the total time available

14 Ko (2017).
15 Rarick (2007).
16 Parkinson (1957).

depends on the number of employees, and the number of employees is dependent on many other factors than the total amount of work: for instance, employment, promotion, and retirement strategies, which are all discussed by Parkinson. Where resources are available to employ people, and responsibility for decision-making is distributed in a system, it is easy to rationalize the multiplication of resources given to the administrative task. Parkinson's observations have been further justified by more recent research.[17] However, the key condition for Parkinson's Law not noted by Parkinson is that in addition to lapse of time and the availability of resources, the system is reliant on the judgement of those who do not identify with their efficient use, because they do not identify sufficiently with the organization's interests as a whole when its goals conflict with their personal goals.

Such observations on the inefficiency of bureaucracy have tended to fuel a backlash since the 1980s. As Donald Savoie has written:

> *By the early 1980's, public bureaucracies stood accused of many things: being bloated, cumbersome, uncreative, lethargic, and insensitive. This widely negative perception could be found in many countries.*[18]

Although recognition of the limitations of bureaucracy was found across the political spectrum, the backlash is particularly associated with neoliberal 'minimal state' ideology.[19] Margaret Thatcher in the UK, Ronald Reagan in the US, and Brian Mulroney in Canada all undertook campaigns to cut down the size of the civil service. However, this approach is typical of many reactions to absolutized beliefs, in failing to question the wider framing of those beliefs. Whilst the top-down organization of bureaucracy remains, and the key cause of waste is believed to be excessive resources, rather than alienation or psychological conflict, administrators within that system have little incentive to counteract it. Merely reducing the resources given to the system (or putting pressure on it through idealistic target-setting) is likely to adversely impact its functioning. As Savoie records, the reality of administrative reform in the 1980s rarely matched the rhetoric.[20] An administrative system starved of

---

17 Klimek, Hanel, & Thurner (2009).
18 Savoie (1994) p. 3.
19 Ibid. pp. 3–4.
20 Ibid. ch. 4.

resources is also less well prepared for new conditions, as many of its 'inefficiencies' provide means for change.

If anything, the attempt to closely control the efficiency of an administrative system results in ever stronger reinforcing feedback loops of *managerialism*: more rules, more records, and closer supervision to bureaucratize the bureaucrats. As Thomas Klikauer argues, this often involves the cloaking of a strong ideological orientation, often allied to a profit motive, as the mere application of management techniques.[21] In effect this also involves the multiplication of the absolutizing bureaucratic mindset in the guise of its avoidance.

The reaction to state bureaucracy continues to be part of the anti-state rhetoric of sections of the political right, but very often the attempt to act on this rhetoric merely makes state administration ineffective, and thus contributes to transferring some of the powers of the state to private organizations, including for-profit companies. This misses the necessary importance of administration in society, and does not even necessarily make it less bureaucratic, as companies can also develop all the same tendencies to 'bloated' bureaucracy as state administrators. The political debate about this tends to take for granted the bureaucratic hijacking of administration, and ignore the psychological aspects of the systemic processes involved, to focus only on the socio-political ones. However, in the wider view, administration is just another of the cultural institutions I have been tracing, which starts off in a provisional form but later gets rigidified. The solutions to this absolutization, as elsewhere, do not lie in more intense management, but in integrative practice.

---

21   Klikauer (2015).

3.k. Relativism and Breakdown

> *Summary*
>
> Relativism is a recently developed and sustained form of negative absolutization, that combines several dogmas, whilst confusing the particularity and incrementality that we need to recognize with equality of justification. Its most common naturalist and subjectivist forms began in the eighteenth and nineteenth centuries, but have become much more widespread in response to globalization. The naturalistic version can produce narrow utilitarianism, but the subjectivist version is associated with the relativistic breakdown of practically vital consensus on issues such as climate change.

To close this section on the lapses of provisional cultural elements into absolutizing ones, I want to comment not on another cultural process, but on an underlying philosophical aspect of the move into modernity – that of relativism. I could, of course, continue discussing many other cultural areas, including, say, education, health, the arts, and political constitutions and practices. However, many of these other areas are discussed in the final section in relation to practices, and also often in more detail later in this series.

I mention the historical aspects of relativism here to connect this section to a theme discussed closely in *The Five Principles*: that is, the ways that we need to beware of absolutizing *from two sides*, and the ways that these two sides interact defensively against the Middle Way as a possible third alternative. There, I discussed the phenomena of sceptical slippage, appropriation, lumping, and unholy alliances as outcomes of the binary nature of absolutization when understood from a consistently agnostic standpoint.[1]

Considered historically, though, the two kinds of absolutization tend to have an asymmetrical chronology, with positive absolutizations achieving cultural dominance earlier than the negative ones that contradict them. Whilst a negative absolutization can happen whenever anyone contradicts a positive one (which must in principle mean that the two are of nearly equal age), the rise of *relativism* as a sustained cultural manifestation of negative absolutization is largely a modern development. I will need to spend a little time

1   II.4.

ensuring philosophical clarity on the term before returning to the historical aspects.

By 'relativism', I mean broadly *the belief that that any one view is as justified as another*, whether the views being compared are those of individuals, or of groups, or even of whole cultures or traditions. Such a belief is counter-dependent on positive absolutization, because it is justified by the absence of any absolutely justified belief, and shares with it the assumption that absolute beliefs are required to justify provisional ones. It is thus justified by sceptical arguments taken out of a practical context and interpreted absolutely, as showing that positive absolutes are 'false'. Thus, for instance, scepticism about a traditional moral rule like the prohibition of sex before marriage is taken to imply that our moral decisions about it are merely a matter of 'personal preference' – that there are no relevant moral considerations as to whether we should have sex before marriage at all, because there are no absolute ones.

Relativism is often confused with incrementality on the one hand and with particularism on the other, but implies neither and is implied by neither. It is not the belief that our beliefs are justified 'relatively' in the sense of being a matter of degree (incrementality). Nor is it the belief that our beliefs are justified only in a particular context of judgement, with no necessary extension beyond that: our embodied position requires that, so it is an aspect of the Middle Way. Standard definitions of 'relativism' often fail to make any distinction between these elements: for instance, the *Oxford Dictionary of Philosophy* calls relativism 'the permanently tempting doctrine that in some areas at least, truth itself is relative to the standpoint of the judging subject',[2] thus focusing on the abstraction of 'truth' but making no distinction between truth and justification, and at the same time leaving the term 'relative' ambiguous between incrementality, particularity, and non-justification. We have to expect all judgements to be made by individuals with a situated and limited perspective, and that these judgements will have a limited justification – if that is what 'relativism' means, then it tells us nothing distinctive, and will only be a surprise to those totally caught up in positive absolutization. The morally important fault of relativism, however, is to confuse this general situation with an equivalence of degrees of justification between all sources, when incrementality,

2 Blackburn (1994).

on the contrary, implies *different* degrees of justification. Indeed, the lack of certainty in our beliefs, that sceptical argument points out, no more implies that any one view is as good as another than that one view is entirely right and the others wrong.

All these confusions can also interfere when we are trying to identify the origins of relativism. Attempts to identify relativism in ancient Indian sources, such as Sanjaya Bellatthaputta (mentioned above in 3.f) are dubious, because there is no clear indication that Sanjaya had a relativist interpretation of scepticism, and the Buddhist accounts on which we are relying may just be straw-manning him to differentiate the Buddha's view.[3] To identify Buddhist accounts themselves with relativism (such as the argument that 'Nagarjuna was a nihilist'[4]) ignores the practical context of the Buddhist interpretation of the Middle Way, and merely shows that a given commentator is stuck in rigid philosophical abstraction. So, though we can't rule out relativism arising in India or elsewhere, the first reasonably clear indication of it is in Greek philosophy – though, again, we are reliant on an account of it from its opponent. In Plato's *Theaetetus,* Socrates summarizes the view of the Sophist Protagoras as 'Man is the measure of all things – of the things that are, that they are; of the things that are not, that they are not.'[5] This explicitly offers the metaphysical over-interpretation of scepticism that is distinctive of relativism: namely that we know from the limitations of our view itself that absolute things do or do not exist according to their appearance. This view appears to have had some influence in ancient Athens, at least, in justifying the Sophist practice of training young men to be effective orators regardless of the values they were orating about.

We have to wait then until modern times for a fully relativist position (as opposed to a mere negative absolute in other respects) to become influential. Modern relativism always seems to depend on selective as well as misinterpreted scepticism. We have a naturalistic version and a wider subjectivist version, as well as the possibility of selective belief in the relativism of 'knowledge', morality, or aesthetics to the exclusion of other areas. The naturalistic version dates back to the eighteenth century, and the subjectivist one to the nineteenth.

3   Ellis (2019) 4.a.
4   E.g. Burton (2001).
5   *Theaetetus* 152a: Plato (1987) p. 30.

Naturalism has its roots in Hume's 'mitigated' scepticism from the eighteenth century, along with his assertion that 'nature' leads us to believe many things.[6] This turns into epistemic determinism amongst his later followers, such as many analytic philosophers, who discount scepticism when they find evidence 'compelling':[7] strangely, the fact that they can reflect on its supposedly 'compelling' nature doesn't seem to make them believe that they have any options in the matter, such is the ideological power of the dogma of determinism. Scientific evidence is thus judged to tell us 'objectively' about facts because of the overwhelming force with which they are said to strike us, but values are separated from this and judged to be 'relative' to the valuer. Since naturalism can give us no prescriptive justification for universal values, we are left with a mere description of the values that we happen to hold due to our cultural or other circumstances. The implications are thus an absolutism of 'facts' but a relativism of values.[8]

The subjectivist or nihilist version, promoted for instance by Nietzsche (despite his theoretical disavowal of it), tells us that 'facts' are just as much 'relative' as 'values' due to our limited perspective, and that our justification is, at best, a matter of heroic creativity.

> Whatever has value in the present world, has not it in itself, by its nature, – nature is always worthless:– but a value was once given to it, bestowed upon it and it was we who gave and bestowed! We only have created the world which is of any account to man![9]

Such creativity of values is, of course, helpful, but we also need to consider the conditions for creativity in wider human experience and in the conditions it engages with. It is no good merely appealing to a particular 'creative' judgement without engaging with those conditions, for that is very likely to leave us in conflict both with our own perspective at other times, and with the perspective of others. Nietzschean relativism thus depends on selective scepticism, not between facts and values, but between one's own heroic values at this point on the one hand and everyone else's facts and values (along with our own at other times) on the other. Without

---

6   Hume (1975) §129 & 130.
7   See Ellis (2017) for my response to one of these.
8   II.1.g; i.1.i.
9   Nietzsche (2006) §301.

some form of integrative approach one may easily end up with relativism despite one's best intentions.

These philosophical versions are formalizations that of course do not necessarily reflect the assumptions of everyone in their societies. Indeed, at the time of their publication, they represented a very small minority view, challenging a mainstream that was positively identified with the absolutizations of Christian theism. However, since then, the relativism they represent has continued to grow in the West, largely fuelled by movement into the ideological stage that enables people to let go of absolute commitment to traditional group values. Whilst the ideological stage can be expressed either through positive absolutes or negative ones such as relativism, the pressures favouring relativism have also grown: exposure to a variety of worldviews and sceptical questions in higher education, globalization (including migration and travel) increasing contact across cultures, and communications technology such as the internet bringing previously separated groups into contact. When our forms of judgement are still strongly ideological but we are brought into contact with so many implicit challenges to positive universals, it is hardly surprising that many people 'flip' into some kind of linked system of negative universals – one of the various kinds of relativism.

The naturalistic version of relativism is thus commonly found amongst those who are scientifically educated, trained to attempt the excision of 'values' from 'facts', despite their inextricable interdependence (every value implies facts and every fact implies values – but we ignore this). As values are then excluded from justificatory investigation, they are often then left unexamined, even if they have actually been absorbed from a scientific or ideological context. One example of the effect of this can be the unreflective application of utilitarian values without consideration of the assumptions this makes about the human capacity to understand conditions. This means in effect that a narrow range of 'facts' is consulted and inconvenient ones are ignored, because there is no awareness of the value of taking into account those further facts. Environmentally and socially destructive mega-projects such as the Three Gorges Dam in China can be justified in this kind of way. Intended to generate large amounts of power and to reduce flooding on the lower Yangtze, the Three Gorges Dam displaced 1.4 million people and caused an ongoing ecological disaster downstream.[10] Yet the ideological justification of this project focuses only on the 'facts' about future benefits, at best weighing them against drawbacks that are pre-selected for their relative weakness, due to the ideological values that are implicitly shaping the whole 'calculation'. The larger question that cannot be asked about such projects is 'Should we be relying on fallible human calculation of effects so much when the negative effects of getting it wrong can be so great?'

In naturalistic relativism there is at least a continuing (at least, theoretical) recognition of the value of consulting scientific evidence. The spread of subjectivist relativism, however, has undermined even this advantage by promoting 'one's own truth' or 'alternative facts'. This seems to reflect an ideological rejection of the possibility of ideological-level thinking by those who still predominantly think in an interpersonal way – so the consensus of their group offers sufficient justification for the negative absolutization of an ideological stance, without any recognition of a need to examine the evidence as a basis of ideological judgement. This has had major effects, as, for instance, it has resulted in substantial rejection of the scientific evidence of climate change. The election of a climate-change-denying US president, Donald Trump, resulted in the withdrawal of the world's biggest greenhouse gas producer from the Paris Climate Agreement. Although reversed by his successor, this has resulted in years of setbacks where urgent action is required.

The relationship between relativism and breakdown should now be becoming clearer. Absolutization in general has produced large sets of reinforcing feedback loops in relation to many cultural areas through history. However, relativism as a specific type of absolutization results in the breakdown of previous consensus over the facts and values to be used as the basis of ideological belief. Whilst we were previously on a path of shared reinforcing feedback loops, then, this results in an acceleration of the process, due to conflict between ideologies adding to those reinforcing feedback loops. If climate change already seemed likely to result in the eventual breakdown of human society, relativism can only boost that process through continual division and re-division of the ideological turf.

---

10 https://www.theguardian.com/environment/2011/may/19/china-three-gorges-dam (accessed 2022).

This is not, however, the hopeless note on which this book will end. We still have a whole section to go, in which we will explore the historical basis of the integrative practices that we can still use to respond to any situation whatsoever – including even this one.

4. A History of Integrative Practices

Some key developments in the practices considered in section 4

Earliest practices	Later practices	Hijacking
Archetypal inspiration: religious experience and symbolism; shamanic healing	Prayer and meditation	Projection/ supernaturalism
	Psychotherapy	Dogmatic theory
Music	Arts	Propaganda, commercialization etc.
Kinship protection and childhood discipline	Ethical observance	Absolutized rules
	Philosophical enquiry → Critical thinking	Metaphysics and absolutized logic
Propositional language and stage development	Meaningful education	'Knowledge' education
	Reflection and autobiography	Dogmatic reinforcement
Humour	Comedy	Ambiguity without reassurance
Stranger hospitality	Travel and language learning	Environmental impact of globalization
Specialization	Recreation	Reinforcing system
Martial training	Bodywork	Idealization

4.a. Archetypal Inspiration

> *Summary*
>
> Archetypal inspiration is probably the earliest and most important integrative practice, because it is needed to motivate other practices. Possible early signs of archetypal symbolism are the use of red ochre on cave walls, and compassionate cannibalism. It then became associated with religious experience, which continued despite the increasing formalization, specialization, and projection of religion through time. 'Secular' arts and concepts also provide archetypal inspiration in the modern period.

The final section of this book reverses the pattern of the third section (the hijacking of provisionality into absolutization) to consider all the ways that absolutization can be turned into provisionality. Although some conditions are obviously better than others for the cultivation of the Middle Way, it does not happen by itself: we need to invest in relatively helpful conditions, where we find them, to extend those conditions and make more provisionality possible in future. This turns reinforcing feedback loops into balancing ones, in which the input of new information helps to modify the cycle, in turn creating new opportunities to use further information. Thus we can continually adapt instead of getting stuck in our judgements – and swept away by conditions.

In the final section of *Five Principles* I have already offered an account of a wide range of integrative practices,[1] so I refer the reader there for a fuller account of how each practice contributes to integration and the Middle Way as a whole. Here, though, I am instead concerned with the *history* of integrative practices. That history can only be a generally encouraging history, because it starts off with ill-adapted patterns of judgement in some respect, and then shows how people of the past and present have managed to change those patterns to a greater or lesser degree. By the end of this section, it should thus be clearer how the cultural form of a *practice* is also an outgrowth of the basic conditions of feedback loops going back to the beginning of life.

The range of chapters in this section (which will have, perforce, to be fairly short) should also reveal how wide the range of ways of addressing new conditions has been, and thus reinforce the message

---

1   II.6.

I wanted to offer in *Five Principles* of how many helpful paths can all be sincere expressions of the Middle Way. However, we will also have to note some appropriations along the way (when these haven't already been discussed), whereby integrative practices can still be turned into disintegrative ones. This continuing danger of hijacking should not obscure the overall hope that integrative practices can give us.

I begin with archetypal inspiration, firstly because it is probably the earliest practice used by human beings, but also because it is a prior requirement for any other practice. To be able to engage in any integrative practice, *inspiration* is needed in the form of a reminder of integrative options: without this, our judgement is constricted to habitual options. These, if we are caught up in a reinforcing cycle, will be binary options: to judge in accordance with absolutized beliefs or (unthinkably) to defy them.[2] I have also previously discussed the *archetypal* nature of that inspiration: that is, inspiration from beyond our current identifications that can help us to engage in one of four basic functions in an integrated way: fulfil goals, avoid threats, relate to the attractive other, and connect with wider potential.[3] All animals do these things, but it is humans, distinctively, who are able to create inspirational reminders that help to maintain their long-term integration in responding to these challenges.

To try to track the history of these archetypal inspirations, then, we need to try to identify symbols used in the past that evidently fulfilled these functions. Of course, the further back into the past we go, the greater the uncertainty about how a symbol was interpreted in a particular context. This uncertainty includes the question of whether the symbol was projected or not. However, as I argued above in 3.c, it is time to cut the twin automatic dogmas adopted by most social scientists when they approach religion: that all ancient archetypal symbols are signs of supernatural belief, and that they had a purely social function. Instead we can see supernatural belief (if it is present – and we seldom have any clear evidence of this) as a projection that is developed from a more basic archetypal function, and the social function of those symbols in binding the community as again a further side-effect of a more basic individual psychological effect. Individuals may be linked in solidarity because they

---

2   I.1.d; II.2.a.
3   Ellis (2022) 1.a.

share the same inspirations (right-hemisphere effect), or they may be *bound* by absolutizations (left-hemisphere effect discussed in 1.i above).

The earliest evidence of symbolic material that may have been used for archetypal inspiration goes back at least to the earliest stages of the paleolithic. The oldest such evidence is red ochre found in South African caves where it has been used by early humans nearly a million years ago.[4] Red ochre can easily be used to mark cave walls, but has no other particular practical purpose, which suggests that its use may have been symbolic, even before the markings created using it start to look at all like figurative art. Its symbolic importance may have particularly focused on the maintenance of relationships, given the resemblance of red ochre to menstrual blood, and the ways this enabled women to signal their continuing fertility to men even when their ovulation was concealed.[5] Red ochre is still used symbolically by the modern hunter-gatherer San group in southwest Africa, where women use red ochre both to decorate their own bodies, and to adorn young men, apparently 'to protect them from accidents before they go out hunting'.[6] Traces of red ochre have also been found on prehistoric 'Venus' statuettes of naked women with exaggerated sexual features.[7] It thus seems possible that red ochre was used from a very early stage as an anima symbol, to inspire men with remembrance of a wider relationship when they were away from home (balancing out the specialization of the hunting context).

Surprisingly enough, cannibalism may be another early sign of archetypal inspiration. For many people, eating the bodies, and particularly the brains, of their loved ones has been closely associated with reminders of a wider emotional context in the face of grief. Beth Conklin's study of the Wari' tribe of Amazonia quotes parents who had lost children as suffering regret that they had been buried, and bemoaning the loss of the earlier practice of eating deceased relatives, which gave them much more comfort, but had been given up during their lifetimes due to outside pressure.[8] For other groups, eating the brains or other body parts of enemies has been associated

---

4   Rossano (2006) p. 351.
5   Power & Aiello (1997).
6   Ibid.
7   Cirotteau, Kerner, & Pincas (2021) p. 21.
8   Conklin (2001).

with their personal qualities (an association that has to precede any desire to have those qualities).[9] Cannibalism as a practice seems to date back half a million years, evidenced by the number of early hominid skulls found split open for eating.[10] Of course, we cannot distinguish at that distance of time whether the motives for it were archetypal, as opposed to merely nutritional or pursuing a desire for revenge, and if archetypal, to what extent the archetype may have been projected into a merely instrumental belief in the magical effects of cannibalism. However, there are obvious reasons for merely nutritional cannibalism being taboo, or at least severely restricted to extreme situations, given the distrust it could sow amongst those who are afraid of being killed for food. Archetypal cannibalism with a compassionate motive in harmony with social solidarity, however, is eminently sustainable, and could have had a valuable social function from the earliest human history.

As already discussed in 3.c above, it is ecstatic experience that provides the likely bridge between these early indicators of archetypal inspiration and the beginnings of what we might recognize as religious archetypes, in the 'upper paleolithic' from about 50,000 years ago. Religious experience, whether or not aided by ingesting plants with psychoactive effects, is likely to have taken place, aided by ritual and symbolism, in deep caves. Early religious symbols, which may have included cave art, could be associated with religious experience to provide reminders of a longer-term wider perspective, whether this was heroic (suggested by hunt scenes in cave art), associated with relationships (suggested by the menstrual connection with red ochre), aiding responses to threats (suggested by the cave art of lions and bears at Chauvet[11]), or more widely associated with the broader potential of the God archetype (as suggested by compassionate cannibalism). In any such archetypal symbols, the clue to their function was not supernatural belief but *contextualization*: if the symbol helped people to place their currently dominant thoughts or emotions in a wider and more sustainable context, they were archetypal. Such contextualization could also be called 'shamanistic healing', particularly if it was helping people to process trauma or grief. Of course, there may have been a wide range of

---

9   E.g. Heap (1967).
10   Rossano (2006) p. 351.
11   Gurevich (2013).

objects with such contextualizing functions that we have no direct evidence of.

As humans developed farming and urbanism, these functions obviously became more formalized. Instead of deep caves, people built temples. Visual art as a source of symbols was supplemented by the written word, which was eventually able to record the mythologies that had provided another crucial element of the archetypal symbolism. The rituals associated with religious experience continued, but gradually became more an expression of social solidarity, or even conformity, and less a direct stimulus to religious experience. The shamans became priests, more skilled in passing on socially prescribed forms than in conducting people into direct archetypal experience. Nevertheless, religious experience and archetypal inspiration remained associated with religion, especially for specialists, such as renunciants or monastics, who tried to maintain the deeper sources of inspiration alongside following the outward social forms.

I will not here trace the detailed development of archetypal forms through and beyond the world's main religious traditions, as I have done this in *Archetypes in Religion and Beyond*. In that book I traced not only the archetypal functions of the most widely used symbols represented in, for instance, Hinduism, Chinese religions, Buddhism, and Christianity, but also discussed how these archetypal functions can continue to some extent even in such contexts as Sunni Islam or iconoclastic forms of Christianity, where visual images of archetypal symbols are forbidden as idolatrous.[12] This requires archetypal symbols to take not a visual but a conceptual form: but since archetypal inspiration requires not any particular form but only an association, concepts can also perform the same function, and often have. I also thus traced the continuance of archetypal function in modern concepts such as nature, truth, rationality, or democracy.[13]

Nor will I here discuss the process of appropriation whereby archetypal inspiration takes us back into reinforcing feedback loops, by being projected into a shortcut version: this has already been discussed in 3.c in relation to religious symbols, and again is more fully covered in *Archetypes in Religion and Beyond*.[14]

12 Ellis (2022) section 5.
13 Ibid. section 6.
14 Ibid. section 2.

What is worth mentioning here to outline the history of archetypal inspiration, however, is the shift from religious to artistic and other forms of inspiration in the modern period. In origin, of course, the arts are not easily separable from religion, which has constantly taken creative expression in visual, literary, and musical forms. However, the appropriation of religion by supernaturalism from ancient times, followed by the Renaissance, in which human inspirations began to shake free of those appropriations, created 'secular' art in which the archetypes could be symbolized directly without them.

Archetypal inspiration is not the only purpose or value of art, which can also be aesthetic (focusing attention), symbolic in other respects, or conceptual. However, many 'non-religious' works of art from the Renaissance onwards have a strong archetypal element. For instance, Beethoven's late string quartets and piano sonatas are probably the nearest thing we have to religious experience expressed in music, engaging the God archetype through sublimity. Renoir's *La Yole* (The Skiff) depicts two female figures in a boat on a wonderfully luminous river – and it is the river that transforms our view of the figures in a way that may provide a sublime anima inspiration.[15] The Romantic movement in the arts, from the late eighteenth century onwards, was particularly responsible for re-engaging the arts with profundity of feeling, and thus often with archetypal symbolic power. Whilst many people now continue to access archetypal inspiration through religion, then, probably just as many do so through the 'secular' arts, and indeed in the many ways that the arts percolate into wider culture.

---

15   Ibid. p. 159.

4.b. Ethical Observance

> *Summary*
>
> Ethical observance may be rooted in kinship feelings, but it is childhood discipline that helps to create basic reflection prior to behaviour. Interpersonal and ideological stages then shape and universalize our moral motives, with the Buddha and Jesus both offering further stretches into more provisional practice. Socio-political values, psychological awareness of inner conflict, and 'culture wars' all form further influencing conditions on our ethical practice, the overall benefit of which is to help address basic sources of conflict.

For those social scientists who assume that ethics is just group binding, it must go back all the way through evolution to the point where our protective instincts for kin began – we feel a very basic, almost instinctual, duty to protect our children, for instance. They may be right about the implicit roots of our motivations, even though their reductionist assumptions doom us to the reinforcing loops of relativism if we take them seriously. We may well have *begun* to have a sense of ethics in relation to our bonds with kin, but that doesn't mean that they stop there.

Ethical *observance*, however, is a practice requiring reflection along with formulation of desirable ethical behaviour, which is then held in tension with actual motives and behaviour. The beginning of this in an individual's life is probably found in the imperial stage, where the motives for reflection primarily involve the avoidance of punishment from parents or other authoritative adults. We might thus expect that such reflections go back to a very early stage of human history, when they were part of a process of individual development. In the process of that development they were refined by the genuine concern for others that might then ensue in the interpersonal stage, and in some cases also by the universalization of the ideological stage.

So, it is impossible to tell when ethical observance began in the sense of individual reflection on conduct. The idea that it develops from some measure of parental discipline, however, suggests a later rather than an earlier date, given that parental attitudes to the upbringing of children in hunter-gatherer societies seem to be highly indulgent by the standards of modern societies. Melvin Konner summarizes the anthropological evidence on this:

> Hunter-gatherer childhood was characterized by close physical contact, maternal primacy in a dense social context, indulgent and responsive infant care, frequent nursing, weaning between two and four years of age, high overall indulgence, multiaged child groups, variable responsibility in childhood, and relatively weak control of adolescent sexuality. These appear to be durable features of the model. Departures from them since the end of the hunting-gathering era constitute a discordance and may have psychological and biological consequences that merit further study.[1]

If we take this as typical of paleolithic life, the implication seems to be a relatively contented society in which psychological development did not need to develop much ethical observance as a means of dealing with social or psychological conflict, because such conflict was not yet sufficiently important. However, we should heed Graeber and Wengrow's warnings about the dangers of over-generalizing about paleolithic societies, and their evidence about the contrasting features that can develop through schismogenesis. So it is possible that some paleolithic societies did have a lot more child disciplining in their culture (there is, indeed, some variability in the evidence surveyed by Konner), and as a result children learned obedience to social rules as the starting point of more complex moral reflection, in the same way required of most children later in history. We will need to leave this point open.

What becomes clearer is that by the time we get to ancient civilizations, and particularly to what Jaspers called the 'Axial Age' of the early development of religious traditions (see 3.c), moral rules are common currency. Merely social rules governing children are generally not written down, but moral rules are, because they apply to adults across a particular group and need to be maintained consistently. Such rules are, initially, often virtually indistinguishable from law, because they emphasize reward and punishment in a way appropriate to the imperial stage. Whether a punishment comes from a person in authority, or a projected supernatural source, in practice makes little difference when each of these continually reinforces the other and builds on a pattern of conditioning going back into childhood. It also makes little difference in practice whether the projected supernatural source is an impersonal system (as in the Indian belief in karma as the basis of cosmic justice) or a divine person – the roots of ethical observance are the same, depending

---

1   Konner (2017) p. 64.

on an identification with future self-interest to suppress or repress more immediate desires.

One thing that is striking about the articulation of ethical rules in the Torah – one of the earliest and most systematic sources of them – is, however, that the threats of punishment and promises of reward are also mixed with appeals to identity. An interpersonal basis of rule-following is developing to at least supplement the imperial one. For instance, the Ten Commandments are prefaced by 'I am the Lord your God who brought you out of Egypt, out of the land of slavery.'[2] Moral rules are being placed in the context of a reminder of the unique identity and shared history of the group.

So far, then, it might seem that, although it allows the development of the individual and adaptation to one's society, ethical observance is not an integrative or provisional practice, but if anything a way for social patterns of behaviour to remain unmodified through the generations. The reflectivity involved is only as much as is required to subordinate the individual to the group, and the danger of repression potentially adds to conflict. However, even in these basic socializing forms, ethical observances do prevent further long-term conflicts – for instance by discouraging violence, stealing, or lying, which are obvious sources of conflict with others. They start off by doing these in a fragile fashion by requiring the opposite ('Thou shalt not...'), but a case can be made that this 'flip' away from absolutizing judgements that are obviously destructive is a necessary part of a development process.

It is only when we get to ideological justifications for ethical observance that an element of provisionality starts to be introduced, as the moral prescription can be checked against its wider motive in ideological belief, even if its roots lie in earlier stages of development. Much of both Christian and Buddhist ethics are ideological in this sense. For instance, the Buddhist five precepts, normally seen as basic training ethics for lay people, are 80% shared with Jainism, and go back to at least the time of the Buddha, in the sixth century BCE. They involve undertaking a practice or training principle of avoiding killing, stealing, sexual misconduct, false speech, and intoxicants. Though rooted very much in basic prudence and social identity, these precepts are also often justified through reference to

---

2 Exodus 20:2.

the Buddhist doctrines of karma and rebirth, and assume a background of individual responsibility for the social order.

Ethical observance at an interindividual level of judgement, though, offers the most fully developed Middle Way practice – undertaken not solely due to social sanctions, group pressure, or ideological conformity (though any of these may, contingently, also be involved), but rather as a way of maintaining the conditions for provisionality. Understood positively (for instance, as the cultivation of love or wisdom), such ethical observance requires much of our mental states and relationships, and is thus inseparable from the path as a whole. However, if we understand it in the more restrictive terms of the negative avoidance of certain types of behaviour, it helps us to lay the groundwork for other types of practice by avoiding the gross disruptions of internal or external conflict, intoxication, or addiction. The Buddhist precepts can easily be interpreted in this way, whatever their ideological background, because of the emphasis on practice (rather than just obedience) in their framing.

Another source that can be interpreted in this way is Jesus as presented in the gospels, especially the Sermon on the Mount in Matthew's gospel. This is the Jesus who tells us not to resist those who wrong us, but to 'turn the other cheek',[3] to love our enemies,[4] and not to worry about food, drink, or clothes.[5] Theologians have long puzzled over these apparently impracticable and idealistic moral teachings, calling them 'counsels of perfection', and most Christians have unsurprisingly long disregarded them in practice whilst continuing to revere them in theory. However, Jesus may well be primarily asking people to open up the options they consider in these situations, rather than offering new impracticable rules to replace the old ones. Of course we have no idea what he 'really meant', and this can be endlessly disputed if one relies only on scholarly criteria. However, the most helpful interpretation seems to be that he wanted to provisionalize people's judgements by making them aware of radically different possibilities from the ones they were stuck in. The ensuing moral judgement would then be their own responsibility, not something specifically or directly prescribed, but the process of judgement itself would be more adequate to the circumstances. It is the very *possibility* of loving our enemies that we

---

3   Matthew 5:39.
4   Matthew 5:44.
5   Matthew 6:25.

most often discount, and the interpretation of Jesus's advice as an impractical command might well just reinforce our tendency to discount it. In relation to the law of his time (which was also the moral rule) he stated 'Do not suppose that I have come to abolish the law and the prophets; I did not come to abolish, but to complete.'[6] The term translated as 'complete' here (πληρωσαι) can also be translated as 'give full meaning'.[7] A fuller meaning, in this case, offers more options and thus more provisionality.

Not surprisingly, though, Jesus's moral advice has rarely been interpreted in such a way. Most often it is interpreted ideologically – and thus selectively. Where the political use of violence is concerned, for instance, the Just War tradition contrasts with occasional Pacifism as the two ideological flips of interpretation of loving our enemies. Even where the socio-political issues are polarized and contested, though, individuals can continue to use ideological ethical observance in a way that may help to maintain some conditions for integration. If, for instance, one has a general sense that one 'should' love one's neighbour, despite the frequent human weakness that leads us to hate our neighbours, this may create a helpful prompt for reflection, even if irregularly and inconsistently. Whilst hypocrisy is a sign of the limitations of ideologically-based moral observance, it is not necessarily a sign of its total failure, and should not be taken absolutely. The same can also be said to an even greater extent for interpersonally-based ethical observance: even if I confine my kindness and non-violence only to members of my own group, it may still prompt reflection that would not otherwise occur, and thus has some benefits as a practice.

The scope and application of moral observance is generally inseparable from the question of socio-political values. As previously noted (in 3.e), the move from interpersonal to ideological ways of judging tends to be accompanied by increasing emphasis on the universal values – justice, care, and freedom – as opposed to the group-binding ones of loyalty, authority, and purity. This shift helps to explain how the modern period generally offers a gradual extension of the scope of moral observance, accompanied by liberalization, globalization, and the development of human rights ideology. As people have engaged increasingly in higher education or

---

6 Matthew 5:17.
7 See Ellis (2018) pp. 92ff for a fuller argument.

professional life, they have increasingly assumed that moral observance involves consideration not only to fellow men of the same race and class, but also to men of other classes, women, those of other races, children, and animals. This is what Steven Pinker calls the 'expanding circle', which he thinks has done much to reduce violence and exploitation during the course of human history.[8] For instance, the moral observance of refraining from killing now increasingly not only excludes killing others of one's own race and class, but also not killing slaves (once considered entirely disposable property), refraining from honour killing of women, not committing infanticide, and (for an increasing number of people in the West) not contributing unnecessarily to the breeding and killing of animals by eating meat. Expanding one's own individual 'circle', from whatever starting point, may be an important aspect of moral observance, reducing external conflict of many kinds and thus the reinforcing feedback loops that arise from it.

At the same time, psychology has also extended awareness of inner conflict,[9] and thus of the possibility that ethics is not solely about how we treat others, but also about how we treat ourselves. Serious addictions – to recreational drugs, alcohol, tobacco, etc. – have become medically recognized as damaging to health during the course of the modern period, which has brought the medical profession more directly into the discussion of our moral attitudes towards ourselves. It's thus increasingly clear that those who define morality restrictively as only about our relationships to others are ignoring the constant interdependence of how we treat ourselves and how we treat others. Our treatment of ourselves in practice nearly always affects others: something that the families of alcoholics, for instance, are well aware of. Unfortunately this point is still neglected by many conventional Christians, who interpret 'love your neighbour as yourself' as a cue for unsustainable self-denial, rather than equity in considering the interests of all.[10]

These new factors of course make moral observance in the modern world increasingly demanding. In order not to simply react against these demands, the Middle Way (at least in some implicit form)

---

8   Pinker (2011) pp. 832-4 & 456-579.
9   I.5.a.
10   Schlabach (1996), for instance, recognizes the unsustainability of the 'self-denying' Christian ethic, and unsuccessfully tries to square the circle by appealing to the projected transcendental view of God.

becomes crucially important, in the sense that we are only likely to start to meet these many demands with enough understanding and acceptance of the actual capacities we begin with, and the incremental stretching of those capacities. This 'stretching' account of morality, and the distinctive approach that the Middle Way can offer to moral interpretation, will be the theme of a later volume of this series.[11] In the absence of any such process, fragile reactions are unsurprising, and we increasingly see these in the form of political reactions against 'political correctness' and 'wokeness' in the early twenty-first century. The danger of such reactions is that they lead us to abandon moral observance as a practice altogether because of a feeling that it has overstretched our personal capacities. The very framing of a personal practice (of, for instance, using non-discriminatory terminology) as 'political' enables it to be projected onto authorities and thus absolutized one way or another, rather than being treated as a matter of responsibility. However, maintaining ethical observance as a *practice* and continuing to see it within that frame is central to its success.

11   VIII.

4.c. Prayer and Meditation

> *Summary*
>
> Contemplative prayer involves openness to inspirational states, and is thus probably as old as archetypal inspiration. Meditation systematizes working with mental states to produce more openness, particularly by stilling stress loops through body awareness. 'Directive' forms of meditation, like prayer, carry the danger of projection, so probably need grounding in 'non-directive' forms such as mindfulness. Attachment to *jhana* states can also create reinforcing feedback for modern practitioners. The Middle Way is both applied to, and evident through, meditation.

For some, prayer and meditation are totally different things. However, this depends on the model one uses to understand prayer: is it a petition to a supernatural entity, or a state of openness to an archetypal inspiration? I would make a clear distinction between petitionary prayer, which involves projection of the God archetype, and contemplative prayer, which does not necessarily involve such projection and can be an integrative practice.[1] The distinction rests on whether one expects a God or god to act independently of one's own experience, or ultimately recognizes one's own responsibility for the process. The Catholic mystic Richard Rohr captures well what contemplative prayer can be like as an integrative practice:

> *Prayer isn't about changing God but being willing to let God change us. Prayer is sitting in the Silence until it silences us, choosing gratitude until we are grateful, and praising God until we ourselves are an act of praise.*[2]

Such contemplative prayer may well precede meditation as a practice, if only because the latter focuses more explicitly and technically on working with one's mental state, even if the goal of this is to engage with the inspiration of the God archetype. Meditation is a kind of technology – not one involving a constructed external object, but nevertheless a practical *technique*, involving enough critical awareness to systematically direct both the object and the mode of attention.[3] Such awareness may require a more or less ideological level of judgement. Nevertheless, one can engage in a prayerful openness without any such technique, just as one may reach

---

1 II.vi.c.
2 Rohr (2016).
3 Eifring (2016) pp. 5–7.

'meditative' states spontaneously as a result of the conditions and one's response to them.

The earliest archetypal inspirations, already discussed above in 4.a, may be seen as equivalent to prayer as a contemplative practice. Of course, we do not know what mental states or attitudes were dominant for paleolithic people who may first have had religious experiences stimulated by psychoactive plants or fungi, and associated these with sacred symbols – perhaps with deep cave paintings where rituals took place. However, if their attitude was wonderment, full of awe for the meaningful experiences opened to them in this way, their attitude could well have resembled that captured by Richard Rohr (as opposed to that of a church congregant who asks God to heal his aunt or help him pass his exams). Just as we have no business to assume that their activities were motivated by 'religious beliefs' in supernatural forces, as I've already argued (even though we can also never rule that out), likewise we have no reason to assume that early human practices were narrowly focused on the fulfilment of goals, rather than on the more open mental states of contemplative prayer. In the paleolithic cave, there is most unlikely to have been any distinction between calling on a god, being in the presence of a god, or being a god. The 'god' was rather a term for the whole set of inspirational associations engaged with in the ritual and practice.

For the development of meditation as a technique, though, the evidence seems to point to India around, or perhaps somewhat prior to, the sixth century BCE: a period of rich religious development. Around that time was the development of a number of renunciant (*shravaka*) groups challenging Brahminical Hinduism, along with the Upanishadic tradition in Hinduism which tries to reconcile Brahminical dominance with the techniques and debates used by the renunciants. These two parallel religious cultures both employed meditative techniques, and evidently stimulated each other through the process of rivalry or *schismogenesis* that Graeber and Wengrow identify.

We cannot tell who first developed meditative techniques, but only that they were used in rather different socio-political ways by the two groups. For the renunciants, which became the basis for Buddhism and Jainism, meditation provided access to a common human integrative practice, and thus helped to challenge the absolutization of Brahminical class-distinction and its attendant purity beliefs. For the Upanishadic sages, on the other hand, meditation provided a further resource for the reinforcement of Brahminical status, using techniques that were later detailed in the *Yoga Sutras* of Patanjali.[4] These divergent ways of using meditation have been reflected throughout its history – with, for instance, the Japanese samurai (or even the modern US military) using meditation to enhance their elite performance,[5] whilst the Ambedkarite Buddhist movement in India uses meditation to help empower oppressed and disadvantaged dalit people.[6]

The Buddhist tradition has developed a particularly rich meditative practice, in which different schools, cultures, and types of technique have constantly interacted, as Buddhism spread from India through eastern and southeastern Asia. The Buddhist tradition distinguishes two different types of meditation: *samatha* and *vipassana*, broadly 'concentration' and 'insight'. The distinction helps to capture two different functions in meditation: namely focusing of attention and awareness of (primarily internal) conditions. However, the definitions and the boundaries between these two types are often unclear or contested even within Buddhist tradition.[7] The key issue seems to be whether *vipassana* involves awareness of conditions that merely emerge from open monitoring of one's experience (as in *zazen* meditation, or 'just sitting'), or whether it is a method of appreciating the 'truth' of Buddhist metaphysical teachings, such as the supposed omnipresence of impermanence. If *vipassana* is pursued in the former more open fashion, it becomes difficult to distinguish from *samatha*, as in order to monitor our experience openly we still have to remain sufficiently focused, requiring a process of relaxation and body awareness balanced with a consistent intention. The value of this kind of practice for balancing feedback should be clear, as it helps us both to avoid stressors that create reinforcing feedback (in the form of mental proliferation or rumination), and to create increasingly better immediate conditions for developing awareness of our thoughts, emotions, and bodily state. However, *vipassana* in the sense of an attempt to digest pre-formed dogmatic lessons from traditional formulation seems in much more danger

4   Bryant (2016).
5   Benesch (2016).
6   Hennigar (2022).
7   Eifring (2016) p. 27.

of creating reinforcing feedback loops through the entrenchment of metaphysical beliefs.

To clarify the *samatha-vipassana* distinction (in the context of a cross-traditional survey of many strands of meditation), Halvor Eifring proposes a distinction between 'directive' and 'non-directive' meditation in its place.[8] Directive meditation, he says, is an 'outside-in' process

> ...as when the meditative effect of mantras comes from their place within the cosmologies surrounding them that endow them with symbolic if not literal meaning, and thus help meditators to 'discover' the knowledge already cultivated by their traditions.

This is *vipassana* at its most dogmatic, for example incorporating claims about the cosmological significance of mantras that supposedly capture the vibrations inherent in the universe, whether found in the Buddhist or Hindu traditions. The meditator then internally chants the mantra and thus supposedly participates in the cosmic vibrations. This 'directive' approach can also fit some Christian approaches to meditation, which likewise aim primarily to reinforce absolute belief, by recruiting experience to reinforce the models that it is pinned on. The meditative approach in such practices is 'concentrative', and the focus of meditation 'thematic', according to Eifring, in the sense that it includes specific concepts, or other symbols closely associated with those concepts: 'a sacred text or image, a holy person, a deity, a moral virtue, an idea, or concept'.[9] Here, then, the effects of the practice are likely to depend entirely on the provisionality of that context. One can visualize a deity, for instance, or recite the name of God, in a way that is framed by archetypal inspiration and personal responsibility for a practice, or alternatively it can have many of the functions of petitionary prayer, reinforcing a metaphysical belief in the separate 'existence' of a supernatural force or cosmological process that substitutes for practical responsibility. The concentrative aspect of directive meditation also makes it open to negative psychological effects, as the practitioner may simply increase conflict by trying to force one perspective on a recalcitrant experience. Some of the worst effects of 'meditation' pursued in this alienated way can actually be to exacerbate rather than address mental health issues.

8   Ibid. pp. 28–9.
9   Ibid. p. 35.

One of the key ways to make directive or thematic practice less projective must be to connect it sufficiently with what Eifring calls 'non-directive' or 'inside-out' meditation.[10] This is much more focused on the mind and body of the meditator. It is rooted in bodily awareness and relaxation, and typically focused on a non-conceptual 'object' such as the breath. A non-conceptual focus allows an integration of concentration (or one-pointedness of focus) with a simultaneous reduction in the likelihood of proliferation, because we are activating the task-positive network of the brain but not subordinating this to the representational centres of the left pre-frontal cortex. This allows us to maintain and develop awareness of wider processes within our experience without it being drowned out by a dominant cycle of representational beliefs. This is 'mindfulness', which has been associated with both *samatha* and *vipassana* approaches in the Buddhist tradition, and which I discussed in a number of places in *The Five Principles*.[11]

In some senses, the two forms of meditation are interdependent: we need a 'directive' element in the sense of an overall intention to meditate by developing a process of awareness, and to do this we need one-pointed concentration. This 'directive' element needs to constantly interact with a 'non-directive' element of openness to new experience, so as to enable balancing feedback loops. The balancing between the two is an aspect of the Middle Way as applied during the course of meditation – also called 'balanced effort'.[12] We need to avoid the absolutization either of the belief that we need to concentrate on the meditation object (turning into 'wilfulness'), or the belief that we need to be open to all new experience (which might quickly lead us back into proliferating cycles where conceptual beliefs take over again).

However, this doesn't mean that we need to 'balance' non-directive meditation with the absolute assumptions that are often used in the many directive practices in the world's religious traditions. These assumptions are not experiential, and will create more conflict that undermines the practice if taken literally. Successful mystics in the Christian tradition, for instance, have managed to focus on the *experience* of God in their prayer and contemplation largely by neither dwelling on nor denying Christian dogmas, but

10 Ibid. pp. 28–9.
11 II.1.d, 2.b & c, 3.f, 4.d, 5.d, 6.c.
12 Ellis (2019) pp. 177–9.

rather shifting their focus decisively towards experience. I have made a more detailed account of Christian mysticism elsewhere, in which I identified various ways in which Christian mystical experience (that is, prayerful and meditative experience) can help to modify outlooks in a more integrative direction, even in the context of a dogmatically-interpreted tradition. These include agnostic views of God, a positive view of the body, a cultivation of integrated states, integration of love with wisdom, identification of God or Christ with the self, a practice of humility, and the nurturing of creativity.[13] Many Christians have maintained some of these integrative features in their practice of prayer and meditation, without it being consistently integrative, because the importing of absolute assumptions that were culturally entrenched in their tradition also continued to influence it. Thus we end up with a very mixed picture. St Bernard of Clairvaux in the fourteenth century, for instance, managed to maintain a strong mystical practice in connection with the image of the Virgin Mary, but at the same time took a leading part in preaching the Second Crusade.[14]

Although Buddhist techniques of meditation were developed much more effectively than Christian ones, through most of Buddhist history they have only been practised by a small minority. In Theravada Buddhism, for instance, meditation has been traditionally practised not just only amongst the monastic elite, but only amongst a small section of that elite – the 'forest monks' as opposed to the 'village monks'.[15] Such patterns changed during the twentieth century, as meditation methods were transmitted to the West: at first by a few more open-minded Western scholars and religious syncretists, later by the first Western Buddhists. Meditation began to be seen as a practice for all, and eventually this popularization also began to influence Asian Buddhism, creating more interest in meditation amongst traditionally Buddhist lay-people.

By the early twenty-first century, mindfulness meditation also began to be popularized beyond Buddhism, used as a separable therapy for general relaxation and well-being, or as an intervention to help people cope with stress, depression, or pain. Some Buddhists have become critical of these developments, which they have labelled 'Mcmindfulness', because of their separation from

13 Ellis (2018) pp. 163–73.
14 Ellis (2022) 5.j.
15 Gombrich (1988) pp. 152–3 & 156–7.

wider Buddhist integrative practice, and thus the likelihood that they will do little to change longer-term individual or socio-political conditions.[16] These criticisms help to identify the ways that mindfulness, like any other practice, can be absolutized, either as a panacea or as an instrument: however, this is not a new development – only the appropriating ideologies have changed. A practice that is helpful because of the way that it provides new optionality to a given individual at a given time, can also become an object of absolutized conceptual belief as the panacea for all people at all times. It can also be used in narrowly directive ways by doctors, trainers, and managers as well as by religious leaders, the Middle Way making it clearer that both types of absolutization can be equally damaging.

Another way that meditation can be appropriated and absolutized is by individuals over-concerned with meditative states or meditative results. These might be absolutized by being interpreted as revelations, even though there is no aspect of a highly open, integrated, insightful, or ecstatic experience, however enormously meaningful, that makes it necessarily infinite or absolute in origin.[17] It makes no difference whether the supposed source of revelation is personal (a god) or impersonal (forces of the universe) in this respect.

Meditation can also produce temporary integrated states that the Buddhist tradition has classified as *jhanas*, based on descriptions in the Pali Canon[18] that may go back to the time of the Buddha or before. As argued in *The Five Principles*, temporary integration states only abate absolutization as long as particular temporary mental states last, but they have great potential value for longer-term integration if *invested* in practice that engages with beliefs in the longer term.[19] Even if meditators don't interpret jhana states as revelatory, however, they may become over-attached to them as goals of meditation, leading to an inner over-emphasis on goal-driven directiveness. This can create a type of reinforcing feedback loop also known as the *hedonic treadmill*,[20] in which wilful effort exerted to reach a particular pleasant experience actually makes it harder to achieve, maintain, or benefit from that experience. Although the hedonic

16   Purser (2019).
17   Ellis (2022) 4.f.
18   *Majjhima Nikaya* 26.34–42 (and a number of other places): Ñanamoli & Bodhi (1995) pp. 267–8.
19   II.5.f.
20   Fujita & Diener (2005).

treadmill can occur in relation to any pleasant experience, the ways in which developing *jhana* itself depends on balanced practice make meditation particularly vulnerable to it. Perhaps this hedonic treadmill is a particular trap for modern practitioners who have become prey to it elsewhere, for instance in relation to technological devices (as discussed in 3.h).

Prayer and meditation, then, offer extraordinarily direct and liberating practices that can help us to change the basic conditions in which we make judgement. Meditation, particularly, is perhaps the *most* direct practice for changing reinforcing loops of judgement into balancing ones. However, the directness of both practices is also their weakness, making them both prone to absolutizing interpretations and temporary in their effects. The techniques and accessibility of meditation (unlike prayer, which lacks such technique) have also continued to develop through history. These developments mean that meditation is now increasingly in a position to provide integrative practice to a wider range of people – used, for instance, in healthcare, in schools, and in workplaces as well as through individual initiative as part of a path.

To these advantages, finally, we should also add the epistemological value of meditation in helping people to experience the Middle Way in action, and thus to help them understand it directly. The Middle Way in meditation was nicely summarized by the Buddha in his metaphor of the lute strings: when we meditate, our strings should neither be too taut, nor too slack, to produce the right note.[21] Either tautness or slackness are the result of different kinds of absolutizing assumption, and the wonderful thing about meditation is that it gives us *immediate* feedback when we slip into either of those extremes, as we then loop away from the practice and regain awareness sometime later, recognizing that we have become 'distracted'. We then try again, and again, gradually reinforcing the habit of seeking, and eventually finding, that balancing point in direct experience. Those who have not had that experience in some form are likely to struggle to understand the Middle Way, and if there is any kind of awakening to the Middle Way in the contemporary situation, we can attribute a lot of the underlying reasons for this to the spread of mindfulness practice.

---

21   Ellis (2019) pp. 105–7; *Anguttara Nikaya* 6.55: Nyanaponika & Bodhi (1999) p. 168.

4.d. Bodywork

> *Summary*
>
> Although sport and dance can also have integrative value, the two main traditions of integrative bodywork are the Indian yoga tradition and the Chinese martial arts tradition. Yoga as understood in the West is largely a twentieth-century synthesis, prone to metaphysical inflation of its origins, but this doesn't affect its practical value. Martial arts develop a quite ancient aspect of Chinese fighting that seeks balancing feedback loops and the Middle Way in conflict.

'Bodywork' is a general label that can cover a variety of embodied practices with an integrative effect. To some extent, any physical recreation or sport can be integrative in the sense that relaxing overspecialization of activity (including too much merely sitting activity) helps repressed awareness to emerge: I will say more about this in 4.k below. The physical movement in dance can also combine recreation with an artistic type of integrative value: for more on that see 4.e below. The 'bodywork' I have in mind here, however, is a more embodied form of meditation, including ritualized movement that may be anywhere on a spectrum from very slow with long pauses to relatively fast. These movements are likely to have physiological benefits when regularly practised, but their integrative value depends mainly on the ways that they help to train the awareness, anchored directly to bodily meaning. In this sense they work rather like meditation, likely to bring us to a temporary state of greater integration that we can then invest in more direct work with our beliefs. In some forms, bodywork practices can also include physical contact between different people using the stimulus of touch (as in massage), but, although this may contribute to integrative body awareness, it can also be seen as a type of therapy (see 4.m below).

The development of bodywork techniques follows two evident cultural paths: one as an aspect of the practice of yoga in India, and another as an aspect of martial arts in China. In Europe, there is a gymnastic tradition going back to ancient Greece, but nowhere near the same emphasis on integrated practice combining bodily and mental awareness that is found further east – at least until the Eastern forms started to influence Western culture more recently. Both of the Eastern strands of bodywork have somewhat complex histories, but they provided the cultural seeds for a modern flowering of many modifications of these two traditions, plus the development of many new forms beyond them.

The origins of 'yoga' (originally meaning 'yoking') in India are subject to a good deal of confusion, because the term 'yoga' is used for a range of meditative, devotional, and even magical, practices to which the bodywork element was a relatively insignificant late addition. It's common for yoga practitioners to claim that yoga goes back to the Indus Valley civilization, because seals have been found there bearing images of figures seated in the lotus posture.[1] This may, at a stretch, perhaps count as evidence of some sort of meditation practice (though, of course, there could be other possible purposes for sitting in a lotus posture), but they are hardly evidence of bodywork. The *Yoga Sutras* of Patanjali, mentioned in the previous chapter, also do not contain 'yoga' in the modern bodywork sense. As David White writes:

> In fact, the yoga that is taught and practiced today has very little in common with the yoga of the Yoga Sutras and other ancient yoga treatises. Nearly all of our popular assumptions about yoga theory date from the past 150 years, and very few modern-day practices date from before the twelfth century.[2]

Bodywork yoga originated in the *hatha yoga* that was first detailed in texts of the tenth and eleventh century CE. This included breath control techniques and the use of stretching postures, but the aim of this was not integrative, but rather the diversion of powerful bodily fluids, the normal tendency of which was to flow downwards, to instead rise upwards.[3] There may have been some element of bodily experience here in the relationship between the *kundalini* rising and the embodied experience of energy rising in the spine, and the envisaging of the *kundalini* as *Shakti*, the goddess, may have had an archetypal function.[4] Nevertheless, the overall framing is ascetic rather than integrative, even apparently when the same framework was adapted from Tantric Hinduism into Tantric Buddhism.

In the late nineteenth century, the *hatha yoga* tradition was revived by Swami Vivekananda, but he did not do much to actually revive postural practice, which had been largely lost in India. It was, instead, Tirumalai Krishnamacharya, under the patronage of the

1 E.g. Basavaraddi (2015).
2 White (2012) p. 2.
3 Ibid. p. 16.
4 Ellis (2022) pp. 209–10.

Maharaja of Mysore, who seems to have been most responsible for developing *hatha yoga* practice in its modern form during the 1920s and 1930s. Krishnamacharya seems to have combined elements of *hatha yoga* with 'British military calisthenics, and the regional gymnastic and wrestling traditions of south western India',[5] and was the teacher of a range of subsequent propagators of yoga – most famously B.K.S. Iyengar.

Modern yoga is thus subject to a lot of spurious justification both through an absolutized tradition that it only partially draws on, and through metaphysical interpretation of the significance of the postures (*asanas*) that are the basis of practice. Here is an example of the latter: 'each movement begins with effort, matures into stretching to reach an ultimate position, then recedes from that to attain balance which is thus a form of transcendence or revelation'.[6] This follows the normal topsy-turviness of metaphysical inflation as I discussed it in *Absolutization*:[7] instead of a spiritual or integrative change emerging from an embodied one, the embodied change is packaged in terms of a projected version of the integrative change. It is the embodied process in modern yoga practice that can be helpful, completely regardless of its origins – the bodily stretching and balancing that helps us to model the stretching of meaning and the balancing of belief, as well as directly helping to create bodily conditions for it. The great thing about embodied practices is that it is quite possible to do them and experience their integrative effects whilst simply ignoring the metaphysical baggage that its teachers might wish to burden you with. Of course, such baggage can also take a negative form – for instance, the claim that yoga is 'only' for relaxation or recreational purposes and has no further value or implications.

For the other strand of Eastern bodywork, we need to start with martial arts – which means a training in fighting skills. Whilst in India, helpful techniques for bodily integration emerged from practices of fighting *oneself* (i.e. asceticism), in China they emerged from fighting outward opponents. Due to the Daoist influence on Chinese thinking about combat, though, it seems that a systemic perspective was available from quite an early stage. If we think of combat as a conflict enmeshed in reinforcing feedback loops (each

5 White (2012) p. 21.
6 Sjoman (1996) p. 45.
7 I.4.f.

side constantly repeating actions that try to eliminate the other, even though they keep stimulating further resistance), then the possibility of engaging balancing feedback loops instead is one that readily presents itself. You can't stop someone else attacking you, but yielding to them completely would often be just as absolute a response as resisting in the same way. Instead, we need to reframe, as some Chinese practitioners have done from a very ancient date, to confront an attacker with the frustration of his aggression – for example by allowing his own strength to be used to frustrate him. This, however, requires considerable dexterity and skill on the part of the defender – a matter not only of bodily training but of immediate mental awareness.

There is evidence that this approach had developed in China by around the fifth century BCE. This is found in the 'Maiden of Yue' story from the *Spring and Autumn Annals* – a collection of brief provincial records spanning many centuries that is also accompanied by various commentaries.[8] This source not only shows the early origins of the martial arts tradition in China, but also the two strands of thinking – 'hard' and 'soft', or 'external' and 'internal'. The 'hard' approach meets force with force, but the 'soft' approach applies dexterity to use the opponent's strength against himself – for instance, by dodging a blow so that the opponent continues further in the same direction where he was expecting to meet resistance, and following that up with a throw that utilizes the opponent's momentum. Stanley Henning, the most influential Western scholar of the history of Chinese martial arts, emphasizes that these two aspects of Chinese martial arts were never completely distinct.[9] Rather the 'soft' or 'internal' approach seems to have been there from the beginning, interacting with the 'hard' one.

It is training in immediate awareness linked to bodily movement, however, that has helped to create the traditions of Chinese and Japanese bodywork that have grown out of this 'soft' aspect of the Chinese tradition. As in the development of modern yoga, we can also credit a creative synthesis here, as Buddhism brought Indian meditative traditions into China and these were incorporated into the martial training. It was in the context of Chinese monasteries that more consistently 'internal' practices have developed. *Tai Ji Quan*

8   Henning (2018) ch. 12.
9   Ibid. ch. 1.

(tai chi) and qigong seem to be the approaches that have evolved to the extent that they might be described as bodywork rather than combat training – although the combat elements are still there, at least symbolically, in many strands. The former tradition is often dated back to the legendary Zhang Senfeng in the twelfth century,[10] but more reliably then to Chen Wangting in the seventeenth century.[11] As these practices have been popularly disseminated both in China and the West, they consist primarily of carefully formalized slow movements – slower and more demanding of attention than dance, but more dynamic than yoga. As Yucheng Guo et al. write,

> *The blending of focused physical activity with breathing exercises in Tai Ji Quan has long been thought to nurture the full integration of body, mind, ethics, and behavior. As Tai Ji Quan involves deliberately executed movements that are slow, continuous, and flowing, it results in calmness, the release of stress and tension, and heightened awareness of the body in relation to its environment.*[12]

Of course, there are also ways in which the balancing feedback loops generally created by this practice may be turned back into reinforcing ones. The dangers of both metaphysical inflation and conventional reduction are present, as with yoga, but perhaps the proximity of tai chi and qigong to martial arts create the most immediate such danger – namely that the skill and dexterity of martial arts training will still be channelled in a way that is largely only competitive. Much depends on how much awareness of the Middle Way as a practice is applied, to determine whether competitive drives are merely being harnessed for integrative ends, or whether they are being reinforced as ends in themselves.

10 König-Weichhardt (2017).
11 Yucheng Guo et al. (2014).
12 Ibid.

4.e. The Arts

> *Summary*
>
> All the major arts have very ancient origins, and offer different kinds of beauty that are all integrative developments of embodied experience: aesthetic, symbolic, archetypal, or conceptual. The performing arts particularly added a socially integrative dimension. After the rigid interlude of the European medieval period, the Western arts were reborn with extraordinary integrative development from the Renaissance onwards. This continues, despite the dangers for the arts of propaganda, commercialization, over-specialization, over-conceptualization, and over-technologization.

The origins of the arts as a practice cannot be separated from those of archetypal inspiration as discussed in 4.a. In order to maintain inspiration that helps us to create balancing feedback loops and adapt to change, humans have used the arts to depict archetypal forms since paleolithic times, this archetypal depiction probably being integrated with religious experience and ritual. Cave art for the most part offers mysterious markings and symbols, but sometimes identifiable humans and animals. It seems likely that archetypal inspiration was by far the most important function of cave art, if not the only one – helping to maintain the heroism of the hunt, the loyalty of relationships, caution against threats, and an openness to new possibilities. However, it is conceivable that cave art was also sometimes used for other types of communication, teaching, or simply imaginative play, all of which are other possible functions of the arts generally.

It seems likely that alongside visual art, music also had a very early place in human history. The patterns of tone that we call melody precede understanding of words in the development of children, and there is a strong case that historically humans were able to use melody and rhythm long before they were able to use language.[1] As Mark Johnson puts it, 'music is the embodied flow of life', with rhythm and melody having direct effects on our bodies in relation to pulse and to the increasing and decreasing of tension.[2] Prior to its participation in language, music seems more likely to

---

1   McGilchrist (2009) pp. 103–5; Dunbar (2004) p. 132.
2   Johnson (2007) pp. 236–42.

have been aesthetic in its function rather than archetypal, meaning that it engaged and focused people's attention, providing a direct sense of beauty rather than primarily an association with other experiences.[3] Beautiful music makes us *listen* – that is, carry on examining and exploring our sensations, rather than prematurely close down our experience with a conceptual summation of the nature of an object in accordance with our beliefs about it. If we imagine an early hominid mother singing to her child (perhaps only with pre-vocal sounds rather than words), the response of the child is to listen and wonder, captivated. It is not a moment for forming conceptual beliefs, or even for associative inspiration, but simply for open attention. That open attention, when we can stimulate it, is one of the most basic integrative practices – the origins of mindfulness. It is the beginning of the most simple of balancing feedback loops – the ones where we carry on looking or listening instead of turning away with a ready conceptual label.

Somewhat later, after the development of language, came storytelling and oral poetry. These arts developed in people's memories and recitation skills long before they were written down. Here, then, was a combination of the kinds of integrative functions that first dominated the visual arts and music – for the literary arts, whether oral or written, combine both aesthetic and symbolic types of function. The aesthetic function is more likely to be important in poetry, where the words have been arranged in a way that adds to their fascination and makes them easier to engage with – prosodic rhythm, rhyme, consonance, assonance, and alliteration are aspects of music imported to poetry to add to the aesthetic attention we give to the words. However, the words also have associations (meanings) both individually and collectively, making them symbolic in effect. We are reminded of other experiences when we hear those words, which help to provide us with new sources of meaning that we can draw on to understand our experience. The story as a whole, though, is likely to have a larger archetypal purpose and effect: we identify with the hero and thus are inspired to heroism; we suffer the hero's fear and thus learn to beware of new threats.

From earliest times, then, the arts had not just one, but multiple forms of beauty that helped to engage us in an integrative process, whether that beauty was aesthetic, symbolic, or archetypal

---

3 VII.1.

(a fourth type, conceptual beauty, required more complex artistic or mathematical forms to develop). These forms of beauty could work together or separately to integrate meaning – either through attention, or development of symbols, or inspiration. There will be a fuller discussion of these forms of beauty and the different integrative functions they serve in a later volume of this series.[4]

The developments of dance and drama have their roots in ritual and storytelling. Dance is present in all cultures. It combines the integrative effects of bodywork (see previous chapter) with aesthetic expressivity, and symbolic or even archetypal power when dance is used for storytelling. The earliest evidence of dance seems to be in Indian cave paintings that may be up to 8000 years old.[5] Drama similarly developed from ritual, where archetypal stories could be re-enacted. By the fifth century BCE in Greece, stories in ritual had developed into theatre, where people impersonated others (stretching our experience of identity) and the format of plays used story to resolve conflict in two different ways: tragedy helped the audience to face up to suffering with empathy, whilst comedy developed the role of humour (see 4.h below) to reframe unnecessary conflicts in ambiguity. Stories became far more powerful when enacted in this way, both because they were accessible and vivid, and because they were shared by the whole audience, creating social as well as individual integration. This Greek legacy influenced the entire later Western tradition, as well as also (probably) the development of Sanskrit drama in India.[6]

The arts flourished in all the ancient civilizations of the world, where they must have played a powerful role in integrating both individuals and society. They did this mainly by adding hugely to the total symbolic resources available, so that people could find new ways to become aware of both their own states and the world around them. These symbolic resources were both 'cognitive' and 'emotional'. People found new ways of saying things, especially from literature and drama, but they also had their attention brought to emotional states through music, visual arts, and dance. This applies to folk forms as much as to the 'high' art patronized by the ruling classes. In ancient Rome, for instance, whilst only upper-class

---

4   VII.
5   Mathpal (1984) p. 214.
6   Keith (1992) pp. 57ff.

men would be educated in Greek literature, everyone went to the theatre.

During the European Middle Ages, the arts continued, but were heavily constricted by Christian ideology. Music, dance, the visual arts, and drama continued to serve archetypal purposes, but were aesthetically, symbolically, and conceptually impoverished in their range and sophistication. Iain McGilchrist attributes this process to a swing of the pendulum back from right- to left-hemisphere dominance. Central to this is the way that creativity changes. Art (unless it is purely abstract) involves imitation of one kind or another, but the hemispheres have different ways of imitating: the left hemisphere through precise, conceptually-led, copying, the right through a *gestalt* process in which one experience in its wider experiential context stimulates another.[7] Whilst the ancient world was awash with creativity of the second kind, in the Middle Ages this is replaced with endless appeals to authority and merely conceptual imitations. McGilchrist describes this as instanced in the visual arts:

> Through the Middle Ages face and body are symbols only: individualistic portraits of the emperors disappear, and they become alike in the same way as the saints. There is a turning away from beauty of proportion, based on the human body; size now represents an idea, the degree of significance we should attach to the figure. Martyrs and ascetics, with their revulsion from the body, replace the classical heroes: all life in the flesh is corrupt.[8]

The visual arts of the Middle Ages start to resemble those of patients with malfunctioning right hemispheres, crudely lacking all contextuality and proportionality,[9] even though these features were previously present in much of the art of the ancient world. In literature and drama, there is a constriction both of themes and models, with the ancient authors selectively referred to as distant authorities.

However, in other parts of the world, the cultural process was completely different. China and Japan maintained a strong tradition of aesthetic cultivation in all the arts. One engaged in them primarily to broaden and inspire through immediate attention. Tibet, on the other hand, eventually developed an extraordinarily rich and complex tradition for archetypal symbology in art. In the Muslim world,

---

7 McGilchrist (2009) ch. 7.
8 Ibid. p. 293.
9 Ibid. pp. 78–9.

there was a cultural flowering between the eighth and fourteenth centuries that showed receptivity to a variety of influences (including ancient Greece and India), with development of the sciences, philosophy, and literary arts. Muslim influence on Europe via the Crusades is one of the stimuli often cited for the beginnings of the European Renaissance in Italy.[10] This does not necessarily mean that these cultures are intrinsically more provisional than the Western,[11] but rather that the pendulum has swung back and forth in different places at different times, thus allowing mutual stimulation to maintain the integrative role of the arts in a global perspective.

The arts in Renaissance Europe, however, were a bellwether of changes later to emerge in science, philosophy, education, and democracy. The development of perspective in painting offers a more embodied version of the 'perspective' that would later develop elsewhere in Western culture – that is, the contextualization. McGilchrist here gives the example of the poet Petrarch climbing a hill to see the view – something people apparently just did not do in the Middle Ages.[12] Thomas Wyatt's concern with his mistress's emotions rather than only his own, and the intense category-defying creativity of Shakespeare, offer further examples.[13] The expansion is largely one of symbol: people just have more to say, more richly, and more means of saying it, with more of their whole embodied experience incorporated into the process. However, that can hardly have been unaccompanied by a new direction of attention: one that we can perhaps glimpse in the aesthetic beauty of Botticelli's sublime angels or Michelangelo's vigorous figures. The themes of Renaissance art also expand dramatically: no longer constricted only to religious themes, Renaissance artists start to draw on the riches of Graeco-Roman classical mythology as an alternative source of archetypal material, then to explore the aesthetics of portraiture and landscape.

As we move forward from the European fifteenth and sixteenth centuries to the seventeenth, eighteenth, and nineteenth, a greater range of aesthetic exploration emerges in painting: Monet's famous

---

10   Trivellato (2010).
11   There is particularly a danger of idealizing oriental culture on the strength of decontextualized evidence of specific integrative approaches at specific times, or of marginal genetic or epigenetic differences (e.g. McGilchrist 2009 pp. 452-9).
12   McGilchrist (2009) p. 299.
13   Ibid. pp. 302-4.

sequence of water-lily paintings would be a strong example of this, because we become interested not in what the water-lilies represent for us, but in their sensual texture in the context of the whole experience of looking at them. Perhaps oddly, this aesthetic focus also develops in the context of twentieth-century abstraction, in the hands of artists like Mark Rothko and Georgia O'Keeffe. The rise of the novel from the eighteenth century is also a major landmark in the development of the literary arts – one that can be credited with helping to create an expansion in sympathy in wider society, as readers began to engage much more closely with what it was like to be someone else.[14] This also roughly coincided with the development of the Romantic movement, which transformed our appreciation of landscape, particularly through poetry. There are also an extraordinary range of developments in Western classical music, from baroque polyphony through to the classical symphony, vastly expanding its expressive range along with the technical capabilities of instruments and players. Although it contains symbolic and archetypal elements, these are generally far less central to the experience of listening to or performing classical music than the aesthetic.

Along with this picture of balancing feedback loops in countless minds, as both artists and their appreciators engage with the flowering arts, there are of course also ways in which the arts have been appropriated and used to fuel reinforcing feedback loops. These can take a variety of forms: support of ideological repression, economic repression through the commercialization of art, overspecialization, over-technologization, over-conceptualization, and relativization.

The use of the arts for propaganda was an obvious feature particularly of Nazi Germany and the Soviet Union, but this draws on the older tradition of religious kitsch – images that project archetypal ideals in a narrow form that over-simplifies their meaning in human experience and thus encourages us to make shortcut absolutizations in our relationship to them. There is a remarkable resemblance of approach between the kitsch Virgin Marys you can buy in St Peter's Square outside the Vatican and the idealized workers of Soviet propaganda posters. Such ideological uses of art are also often similar to those of commercial exploitation: if a producer is only interested in what sells, they are very likely to encourage projections as an

---

14   Pinker (2011) pp. 211ff.

immediate way of binding groups in the purchase of their products, whether they do this in the products themselves or in advertising.

Over-specialization in the arts, on the other hand, can be the product of the Romantic revolt as an over-reaction to ideologized art. 'Art for art's sake' was the rallying cry in the early twentieth century of those who removed art from any wider context and insisted that the artist's role was unique.[15] Any contextualization, however, will probably remind us that art can never be for art's sake, but only for humans' sake, and that the effects of artistic over-specialization are likely to be similar to those of any other kind (as discussed above in 3.i). Seeing art in this over-specialized way is also likely to be similar in effect to relativization – seeing the art as only valuable according to the values in a particular context, where 'beauty is in the eye of the beholder'. If beauty was not somewhat more than that, it would have no role in integration.

Over-conceptualization in modern art is extensively bemoaned by Iain McGilchrist, who traces the similarities between conceptual reductions in modern art and the perspective of schizophrenics whose left-hemisphere dominated outlook reduces everything to alienated conceptual parts.[16] This does identify a problem in some modern art, but I find McGilchrist's account of it very one-sided when it comes to modern art as a whole, which has many other facets. McGilchrist also focuses only on a symbolic type of beauty, and thus ignores the ways that modern art has become integrative in both aesthetic ways (as in Mark Rothko, mentioned above) and conceptual ways. Conceptual beauty is often considered the preserve only of mathematicians, but it can also be found in, for instance, the plotting of detective novels, or the conceptual imagination of the short stories of Jorge Luis Borges.[17] If our attention is engaged by conceptual forms, those forms can help us develop our attention: these only become problematic when they are taken as describing the whole picture.

Over-technologization is perhaps the most tempting trap for the arts at present. It has recently become possible, for instance, to enter a description into an AI programme that will then produce a picture for you. Chat GPT can also write an essay, or even a poem. These works are, of course, complex reproductions rather than gestalt

---

15 VII.2.
16 McGilchrist (2009) pp. 408-21.
17 VII.3.

creations, but they use AI learning to draw on such a wide range of materials that they could easily be mistaken for human creations. These new uses of technology in the arts challenge us to focus, not on the product, but on the process of creation in the human brain. To become totally dependent on technologized art (as has already happened a good deal in the production of popular music) is to repress human creativity and the balancing feedback loops that it can provide, at least for the creator of the art. Though the work can still conceivably be creative for the viewer, the chances of this also fall when it has not passed through a human right hemisphere. However, if we treat technologized art like other technology (see 3.h), and see it as a tool to be put in a wider context of creativity, this danger may pass.

Despite these dangers, the arts continue to be one of the most accessible and valuable practices open to all human beings. Whenever we are trapped in loops – whether these are loops of depression, addiction, or alienation – all we need to do is to reach for the novel, the book of poems, or the musical instrument, or to step inside a gallery – and we may immediately encounter a wider perspective. We are not doomed to one narrow view. We have options. The arts remain one of humankind's greatest achievements in reminding us of that.

4.f. Philosophical Enquiry

> *Summary*
>
> 'Philosophy' throughout its history has included both metaphysical claims that induce reinforcing feedback, and critical enquiry that supports balancing feedback. Socrates and the Buddha helped to establish the dialogical technique of the latter. Ancient scepticism and its later revival helped to stimulate systematic critical enquiry. Christian dogma and modern analytic philosophy have each greatly constricted philosophical enquiry at times, but Hume, phenomenology, pragmatism, and philosophical education have all contributed much to keeping it alive.

'Philosophy' can be generally understood as the examination of our most basic beliefs and assumptions, particularly the ones we are most likely to take for granted. However, what is often referred to as 'philosophy' historically can take two very different forms. On the one hand there is abstract speculation about ultimate 'realities', often asserted with only highly questionable justifications – I would prefer to call this metaphysics[1] rather than 'philosophy', although conventionally metaphysical claims are assumed to be part of philosophy. On the other hand, however, there is a process of *enquiry* about the justification of our beliefs. Whilst metaphysics creates reinforcing feedback loops through circular justification of the same abstract assertions,[2] philosophical enquiry offers balancing feedback loops by obliging us to adjust the assumptions through which we interpret the world – that is, assuming that we pay sufficient attention to it and attempt to apply its insights practically.

As to which of these types of 'philosophy' came first historically, it is probably a chicken-and-egg question. Much philosophical enquiry involves the critique of metaphysical assertions, in response to which metaphysicians desperately attempt to find further justifications for their metaphysics, leading to further critique, and so on. In ancient times, the evidence seems to point to dogmatic metaphysical philosophy coming first, and philosophical enquiry later, but this impression is subject to survivorship bias – it may just be that people were more inclined to record or remember the gnomic utterances of supposedly wise dogmatic philosophers, and

1 I.4.a.
2 I.4.c.

the critiques of their contemporaries are now lost. In Greek philosophy, for instance, what we have left of the work of 'Pre-Socratic' philosophers is largely dogmatic in form – but that is no proof at all that they didn't either engage in some kind of enquiry, or get subjected to it by others. It's generally more important whether the socio-political context encourages enquiry – which probably means both that a critical mass of people have reached an ideological stage of judgement, and that the rulers of the time either tolerate philosophical discussion, or are practically unable to prevent it.

The development of philosophical enquiry evidently depends entirely on the presence of sceptical argument. I have already discussed this in 3.f. It is the cultural presence of sceptical argument that stimulates philosophical enquiry in the context both of the Buddha and of Socrates, by allowing framing assumptions to be questioned. In both of these cases we have what appear to be records of dialogues in which philosophical enquiry took place, dating back to around the fifth century BCE. Socrates stimulated such enquiry by what has become known as 'Socratic questioning', in which the questioner engages with his or her interlocutor not by giving their opinion, but by asking a series of structured questions about the other's opinion. These will typically ask for definitions of terms, and confirmation of apparent assumptions and implications that appear to conflict with the interlocutor's intentions. In this way, Socratic questioning helps to identify conflicts in the interlocutor's beliefs that are in need of integration, giving them a wider context so that they no longer each appear to be the whole story and no longer continue to reinforce themselves. However, rather than isolating these contradictory beliefs with the teacher's authority, Socratic questioning helps the interlocutor to identify them for him- or herself, making it less likely that the new recognition of conflict will be repressed due to pride or a wish to save social face, and gives the interlocutor responsibility for the whole process.

Many examples of this technique in action can be found in Plato's dialogues. For instance, when questioning Euthyphro, who is overconfident that he knows what is 'holy' to such an extent that he can appeal to it to prosecute his own father, Socrates first asks him what holiness means, and gets an answer that is merely focused on the holiness of Euthyphro's recent deeds (rather reminiscent of the modern politician's 'It was the right thing to do'). Socrates then explains that he wants a general definition of holiness so that he can recognize different things that are holy in different circumstances, and continues to insist on the generality of the definition needed. In the process he tries to coach Euthyphro into thinking more universally.[3]

A very similar method can also be found in the Buddha's questioning of his interlocutors in many of the Pali Suttas. Just as Socrates probes Euthyphro on the very idealized concept that he is most concerned with in his law case, the Buddha asks the brahmin Sonadanda, in the discourse of that name, about the qualities of the 'true brahmin' – the very thing he thinks he knows most about. When Sonadanda offers five traditional qualities of the true brahmin, the Buddha then asks which of these qualities are really essential – an approach that, like the questioning of Euthyphro, forces the interlocutor to think universally rather than solely in the terms of his group or current concerns, and does not extend to claims about metaphysical essentiality. The discussion of the 'true brahmin' thus gradually metamorphoses into one of the good person in general.[4]

The history of philosophy from this point, then, can be roughly summarized as a mixture of philosophers making metaphysical claims, and the use of such enquiry to probe those claims. In the case of Aristotle and some other ancient sources, some observations and claims based on experience do also begin to appear, albeit mixed in with metaphysical assumptions, and this early science (discussed above in 3.g) can also be the object of enquiry. In the ancient context, the process of philosophical enquiry can be seen quite directly as an integrative practice, because the results had direct practical implications for those involved. When Socrates questions Euthyphro, it is in the direct context of the question of whether he was right to prosecute his father for 'unholiness' or not, and for Sonadanda, the immediate practical question is how to treat this new great *shramana* teacher who questions all of his group's traditions. The direct relationship between enquiry and practice is particularly evident both in early Buddhism and in the Hellenistic philosophies of the Graeco-Roman cultural world in the centuries following Socrates – Stoicism, Epicureanism, Aristotelianism, Pyrrhonism, and Cynicism. Although these philosophies all maintained some

---

3   *Euthyphro* 4e–6e: Plato (1961). See Ellis (2022) pp. 40–1 for a more detailed discussion of this example.

4   *Sonadanda Sutta, Digha Nikaya* 4: Walshe (1995) pp. 125–32. See Ellis (2019) pp. 74–7 for a more detailed discussion of this example.

metaphysical assumptions,⁵ it is clear that enquiry is an integrative practice for all of them, and in this sense they were unified in a wider Hellenistic 'therapy of desire', as Martha Nussbaum calls it.⁶

From this point in the development of Western philosophy, however, the practical value of philosophical enquiry is rarely foregrounded, and metaphysical assumptions are highly dominant until they begin to be questioned by Hume in the eighteenth century. I do not have space here for a detailed history of the whole of Western philosophy, but it is probably enough here to comment that its *practice* never completely died out as a helpful method of examining assumptions from whatever starting point was imposed by the surrounding cultural conditions, even though the range of beliefs that could be questioned became at times very narrow indeed. One of the low points is the medieval era, where philosophy was generally subordinated to Christian dogma. Almost as low a point has been struck also by Anglo-American analytic philosophy from the mid-twentieth century onwards, where a complete abstraction from practice and experienced value has often been the norm, and increasing specialization hopelessly tries to mimic the sciences. As an anecdotal illustration of the latter, I could mention my visit to an analytic ethics conference as a naïve postgraduate student of philosophy in the late 1990s. When it came to lunch, I had to explain that I was a vegetarian (actually a vegan, but I think I was being cautious). I was met by the other 'ethicists' not with philosophical challenges (which would have been fine), but with incredulity. 'Why are you a vegetarian?' they demanded, in tones that made it clear that this was a departure from the conventional group norm (where nobody should take the idea of *practising* unconventional ethics seriously). It felt a little like a twelfth-century monk questioning God's existence in the refectory.

However, there have also been high points at which new forms of philosophical enquiry have been introduced, allowing new aspects of belief to be examined. The re-discovery of sceptical argument, following a Latin translation of Sextus Empiricus's *Outlines of Pyrrhonism* by Henri Etienne in 1562,⁷ stimulated a wide range of new philosophical enquiry, as already mentioned in 3.f. This

5   See Ellis (2001) 3.e & 4.b on such assumptions in Stoicism and Pyrrhonism.
6   Nussbaum (1994).
7   Popkin (1964). Despite his Latin name, Sextus was Greek and wrote in Greek.

began with Montaigne's adoption of Pyrrhonism,[8] and stimulated philosophers as diverse as Descartes and Hume, who each tried to overcome sceptical argument from opposing viewpoints during the succeeding centuries. Writers like Montaigne (in his *Essays*) and Descartes (in his *Meditations*) increasingly internalized the form of enquiry that Plato had put in dialogue form, so that the sceptical voice was considered and responded to within their own argumentative narrative.

Hume also marks a high point, because of his use of rigorous empirical argument employed much more consistently than in his empiricist predecessors Hobbes and Locke (for instance to arguments about the self and about religion). Although Hume makes dogmatic assumptions of his own, his enquiry questions both religious and rationalist absolutizations of all kinds ('school metaphysics' as he calls it), so is by far the most far-reaching up to that point. Although logical positivists and analytic philosophers from the twentieth century saw themselves as Hume's successors, perhaps those closer to his spirit of enquiry were Kant (who was 'awoken from his dogmatic slumbers' by Hume), the phenomenologists (who turned Hume's empiricism into a more consistent experientialism), and the early American pragmatists like James and Dewey (who first tried to use *practical* experience as a starting point). A more recent stimulus to philosophical enquiry has been added by metamodernism – a still young and far from unified conception, but one that attempts to navigate between the extremes represented by modernism and postmodernism.[9]

One could be forgiven for the superficial impression that all complaints about the constraint of philosophical enquiry in the modern world are straw men, because of the role that philosophical education can still play in opening up discussion for students, often in the best spirit of Socrates or Hume. Philosophical education is the shop window of philosophy as a discipline, and its proponents often still see themselves as practising no-holds-barred Socratic enquiry. However, the reality for those who study philosophy from PhD level onwards is one of over-specialized constraint, as already discussed in 3.f and 3.i, in which the exploration of theories that do not accept current specialized assumptions is in practice professionally

---

8   Paganini (2008).
9   Storm (2021).

blocked by the peer review system, together with specialized publishing requirements for career advancement. This does not stop philosophical enquiry as a practice being to some extent propagated by philosophers through philosophical education, and thus providing some balancing feedback loops through the discovery of new possible perspectives by the few who practice it. However, as often, these balancing feedback loops are also frequently hijacked by reinforcing ones, in a way that I can only find especially ironic in the context of a discipline that prides itself on free and creative thinking.

4.g. Education

> *Summary*
>
> Education is an integrative practice where it supports the development of meaning (including understanding) and justified belief, but merely creates reinforcing loops when trying to inculcate 'knowledge'. It needs to increase in formality to help people negotiate successive psychological stages in widening social settings. Past approaches using rote memorization of texts or facts, or narrow training in skills, can be integrative only as a side-effect. Modern student-centred approaches improve on this, but are often in conflict with the assessment system.

Education, like science, is a process of learning, often aided but never completed by teachers. However, as a result of metaphysical inflation,[1] learning has often been understood as the imparting of 'knowledge'.[2] If this was what education consisted in, it would offer only a reinforcing feedback loop in which existing beliefs in one generation are reproduced in exactly the same form in the next. Instead, education as an integrative practice is a *method* for developing understanding and justified belief (primarily the former). Understanding is not 'knowledge' but meaning – that is, an embodied and associative relationship between symbols and experiences. As we become more educated, we associate an increasing range of symbols with our experience, and combine them with increasing complexity, providing us with greater resources for forming beliefs about the world. To a lesser extent, particularly through the study of philosophy and science, education may also help us to develop more justified beliefs. Education thus aids us both in the integration of meaning and of belief, with the latter dependent on the former, and both aiding balancing feedback loops. However, the idea that this is 'knowledge' is a mark of the *failure* of education as an integrative practice.[3]

---

1   I.4.f.
2   See II.1.a on the question of how knowledge is defined, and why, if it existed, it would be absolute.
3   I have elaborated this view of education in a partly-drafted book on the subject that may be completed and published at some stage. However, it has had to be subordinated for the moment to the books forming the core of Middle Way Philosophy in this series. For the moment, its justification lies only through the consistency of this account with the remainder of Middle Way Philosophy.

'Informal' education, in which children learn new meaning, begins from infancy, and largely happens spontaneously as a child learns language. However, the parents or other 'teachers' have a role, not so much in structuring the experience as in simply providing sufficient stimulus. Early formal education, too, tends to concentrate on helping children to acquire the meaning and generic skills they will need later, from literacy and numeracy through to categorization skills, understanding of basic causal relationships, and manual dexterity. It may involve some demonstration and guiding of these skills, but largely consists of systematically giving children the opportunity to acquire them by providing appropriate framing, facilities, and encouragement. On the one hand, children develop their active involvement with embodied schemas,[4] and on the other, extend their associations with the metaphorical extension[5] of those schemas. This 'informal' process continues throughout our lives, and how 'educated' we are often depends on the extent we have encountered suitable framing, facilities, and encouragement rather than how much we 'know'.

The process of education obviously needs to be understood in relation to the psychological stages discussed in part 2 above. As mentioned there, the transition from the incorporative to the impulsive stage is closely related to the acquisition of propositional language, and the transition from the impulsive to the imperial stage to the ability to develop coherent models of concrete situations by relating propositions to each other. How much the acquisition of linguistic skills is innate or environmental has been the subject of long fruitless dispute, but whatever their respective proportions, linguistic stimulus constantly interacts with genetic ability to help children progress through these stages.

When it comes to the interpersonal and ideological stages, however, wider society has a crucial role. As Robert Kegan notes, the instigators for getting over the tipping points broaden out as we grow up – beginning with mother and family, but developing to peer group for the interpersonal stage, and higher education or profession for the ideological stage.[6] This strongly suggests that the kind of new meaning required to help us continue to adapt as we grow older needs to be increasingly formalized, to ensure that we receive this stimulus when we need it from a wider society that

---

4   IV.1.g.
5   IV.1.h.
6   Kegan (1982).

does not have the obvious stake in our welfare that parents have. The importance of *formal* education to make education in general an integrative practice, then, increases with age, at least up to the ideological stage.

The history of 'formal' education is, however, a varied one in which supporting people to develop new meaning has often conflicted with socially-prescribed metaphysical models. As an institution it seems to have developed in ancient civilizations alongside literacy, as a means whereby those with scribal skills could pass on those skills. Those who were educated in this way in the ancient world – whether in Mesopotamia, India, China, or Greece, for instance, were a small minority, usually only male, whether that minority included the ruling class trained in martial skills or was focused only on a separate clerical class. In ancient Athens, which seems to have pioneered educational development around the fifth century BCE, basic education consisted in reading and writing, gymnastics, music, and poetry. This at least introduces people to some degree of bodywork and the arts as integrative practices, and provides the basic literacy that people would need to build on for more complex critical reflection. However, in that context, only a very small minority of elite young men might go on to study rhetoric and philosophy in the way portrayed in Plato's dialogues.

In many ancient and medieval cultures, however, there was a strong emphasis on rote memorization of texts. We find this for instance in ancient Greece, where one character in Xenophon's *Symposium* describes having to memorize the whole *Iliad* and *Odyssey*,[7] and in monastic education in Buddhist India, where some monks would learn huge sections of the Pali Canon.[8] Such practices continue today, for instance, in the *madrasahs* or Qur'an schools of Muslim tradition, in which children learn passages of the Qur'an in Arabic, with little or no understanding of them, and no exploration of their meaning.[9] Such practices are sometimes defended on the grounds that they provide a fund of remembered text that can later be explored meaningfully by adults, but any such advantages seem hugely outweighed by the conflict they create through alienation from a meaningless task. They also actively discourage

7   Xenophon *Symposium* 4 (Xenophon 1998).
8   Gombrich (1988) pp. 152-3.
9   See Hashim, Rufai, & Nor (2011) for evidence from Malaysia, Indonesia, and Nigeria. I have also observed this personally in a madrasah in the UK.

understanding by deferring it in favour of a literalized adherence to the words of a revered text as absolutized means to an end. We generally remember things *because* we find them meaningful, not vice-versa.[10] If such past education resulted in integration of meaning, then, it was largely as a side-effect, but sufficient of a side-effect to still make a substantial difference. In that respect it was similar to the medical treatment of the time, which included some practices that can now be appreciated as actively harmful (such as blood-letting), others what were neutral, and a few that were beneficial.

Of course, such attitudes to early education are also often likely to be accompanied by dogmatic approaches to belief, so that when children do learn the meaning of what they are reciting, it is likely to be only in the terms of one prescribed meaning overlain with absolute beliefs. We are still dealing with the later fallout from such methods today in the form of the *original language fallacy*[11] in the context of religious traditions – that is, the belief that 'the truth' of a text can only be recognized in its original language, which must thus be laboriously learned, regardless of the costs of doing so. The liberal arts curriculum, associated with Plato and used in schools and universities through the medieval period, concentrated entirely on formal content: grammar, logic, and rhetoric (the trivium) followed by arithmetic, geometry, music, and astronomy (the quadrivium). If one focuses only on grammar rather than the meaning of language, logic rather than critical thinking, and rhetoric rather than justification, the result is an inflation of the significance of those formal connections between assumed metaphysical truths[12] or (in the case of rhetoric) their presentation. The elements of the quadrivium (with the possible exception of musical practice) also focus on formal and disembodied elements of experience at the expense of meaning linked to the body.

When such approaches to education were ultimately overthrown, in a gradual process between the eighteenth and twentieth centuries, it was often to be replaced by a utilitarian approach to pedagogy, which substituted the formalization of meaning with

---

10   This point can be evidenced through prototype theory: we are most likely to recall a prototypical image that most meaningfully stands for a given category: Taylor (2011).

11   iv.3.l; V.3.

12   I.4.f.

the 'useful' inculcation of knowledge. Jeremy Bentham's plans for a fully rationalized utilitarian school for the middle classes, described in his book *Chrestomathia*, encapsulate this approach. The curriculum is now to be dominated by scientific and technological subjects, with the instruction in classics that had dominated in the English grammar school system banished, along with the arts and humanities. To make the teaching process maximally efficient, Bentham also proposed a *panopticon* system of surveillance similar to that designed for his planned prison of that name, allowing one teacher to supervise a thousand pupils. The pupils would get on with their studies because they were always being watched, with a system of reward and punishment to continually modify their behaviour. As Elissa Itzkin points out, this scheme borrowed many features that were actually already operating in schools of the time under the 'Bell-Lancaster' system, but merely took them to their logical conclusion.[13] This built on the 'empty vessels' view of education satirized by Charles Dickens in his novel *Hard Times*, in which children were regarded as so many vessels to be filled with knowledge. A formalistic education in meaning had now been usurped by the apogee of topsy-turvy education in 'knowledge' – still the concept that dominates much thinking about education and its purpose. The role of 'knowledge' as at best the tip of an embodied iceberg was simply not understood because of culturally entrenched absolutization.

The nineteenth and twentieth centuries were, however, accompanied by the gradual development of mass education and the expectation of universal literacy, beginning in Western countries but then spreading around the world. The burgeoning and the globalization of higher education is another aspect of this. This growth has at least provided a large spread of opportunities for integrative development by addressing many of the basic conditions required. However, it has been overwhelmingly driven by economic considerations, both at the level of public policy (where the need for skilled workers is seen as central to economic development) and at the individual level (where the most common motivation for education remains employment opportunities). Even if a Benthamite focus on 'knowledge' has now often been supplemented with one on 'skills', the narrow focus of each remains. They are embodied only to the extent that this enables the fulfilment of certain goals,

---

13   Itzkin (1978).

not in a broader sense. Systems of examinations and qualifications provide a formalized educational currency for those narrow purposes which often bear a very limited relationship to the level of integration achieved by a 'qualified' person.

The most encouraging developments in modern education are those that have recognized the vital importance of embodied and enactive approaches, particularly for young children. Maria Montessori pioneered child-centred learning from around 1906 onwards, finding an effective Middle Way for early education that avoided either inculcating 'knowledge' or complete relativism. Instead, a Montessori School offers a prepared environment, geared towards the needs of a child and modified to meet those needs on the basis of a process of observation, but an environment where children can play freely and make their own independent judgements. It thus harnesses intrinsic motivation and enables children to find and develop what is meaningful to them through active exploration.[14]

Student-centred learning with active approaches has also become increasingly emphasised in higher levels of formal education, at least in teacher training of any kind. At least discussion, rather than just listening and note-taking, is now common practice in higher levels of education: but discussion needs to be intrinsically motivated. Disembodiment and absolutized knowledge are much more deeply entrenched by educational bureaucratization than any student-centred approach can be, particularly when enforced by assessment through examination that tests recall of 'knowledge' rather than the ability to apply a deep generic skill. Where skills are tested, they have to be very specific and measurable. I'm aware from my own experience as a teacher at sixth form (senior high school) level that the power of the means of assessment as the prime motivator for young people constantly undermines all attempts at approaches to learning that do not copy its representation and memorization model. The subtleties of teacher training are often powerless beside the culturally entrenched educational absolutizations that in practice dictate the experience of teachers and students in schools and colleges.

It thus often seems to me extraordinary that modern education makes as much difference to people's lives as it appears to do. Yes,

---

14   Özerem & Kavaz (2013).

there are some students whom it manifestly fails, who end up alienated and in conflict with what they see as the social establishment. However, there are also many students who learn new meaning, test and develop their beliefs, and form at least some critical thinking skills under the influence of education, despite the fact that their education has not particularly focused on these things at all, but more often on 'knowledge' or skills that will be relevant only to a small minority like the method of calculus, the precise elements of the periodic table, or the success of methods used to set up a business. As already discussed in 2.e and f, the role of higher education in transition to the ideological stage of judgement is crucial. I would suggest that much of education's relative success may be attributed to the social role it plays, quite distinct from its actual content. It enables students to work through the stages of development in a way that is increasingly independent of their family backgrounds, with high school particularly bringing students into contact with many of their peers at the time they need to form peer-relationships, and higher education often forcing independence of judgement by separating the student from their original home environment and offering a proliferation of new stimuli. Students often succeed *despite* the manifest failings of their formal education rather than because of them, because their stage development involves balancing feedback loops, even whilst their educational experience is leading them into reinforcing ones.

None of this prevents education from being an integrative practice, particularly in a world where access to its opportunities is increasingly widespread. It can be an integrative practice when we use it to build meaning, not 'knowledge', when we are intrinsically rather than extrinsically motivated, and where we increasingly practise the Middle Way by maintaining awareness of the conditions of learning alongside the 'content' we are trying to learn. Those who are educators or part of the formal educational system can also help to make education integrative by focusing on what I call 'deep generic skills' such as systemic interpretation, navigation of relationships, cultivation of meaning and inspiration, critical awareness of biases and credibility factors, and the use of scientific method. However, these will need elaborating in another context.

4.h. Humour

> *Summary*
>
> Humour seems to have survived unaltered by human history, with unknown origins and a strong presence in ancient Greece and Rome. It works as an integrative practice by integrating meaning that is unsettled by ambiguity, but within a context of reassurance. This works as an account of integrative humour of different types and historical periods. The balancing feedback loops become reinforcing ones when humour is used without a reassuring context, to reinforce power relationships.

The capacity for humour, and thus for laughter, seems to be rooted in the very early history of humankind. We have no specific evidence as to when it began. However, it seems that early European anthropologists encountering Australian Aborigines very quickly noted their capacity for humour. Since the Aborigines are said to have diverged from the genetic tree of the rest of humankind about 35,000 years ago, but still share the same capacity for humour with them, this suggests that humour goes back at least 35,000 years.[1] The first historical evidence for humour comes from ancient Greece, where jesters and jokebooks are recorded,[2] and where the first comedies were produced – for instance in the work of Aristophanes. Not all 'comedy' is humorous, but there is plenty of slapstick in Aristophanes,[3] dating back to the fifth century BCE.

Debate about the function of humour for humans has suggested ways that it is both an innocent way of releasing tensions, and alternatively a mode of repression and the maintenance of social power.[4] As with religion, however, the idea that it could be an integrative practice does not seem to be on academic agendas. Whether or not humour began as an integrative practice, it had developed as such by the time it became enculturated in historically recorded ways. My argument, already briefly begun in *The Five Principles*,[5] is that humour is based on ambiguity, and integrates meaning by unsettling our associations within an overall context of reassurance. Where this context of reassurance is missing, however, and the

---

1 Polimeni & Reiss (2006) p. 348.
2 Ibid.
3 Kapparis, Arnaoutoglou, & Spatharas (2017).
4 Polimeni & Reiss (2006).
5 II.6.e (pp. 245–6).

unsettling of associations is only used to maintain an absolutization in a context of social power, humour has been hijacked and ceases to be an integrative practice.

Let's take an example from a Roman joke book called *Philogelos*, or 'the laughter lover', recently discussed by Mary Beard. Here is one joke:

> A barber, a bald man and an absent-minded professor [are] taking a journey together. They have to camp overnight, so decide to take turns watching the luggage. When it's the barber's turn, he gets bored, so amuses himself by shaving the head of the professor. When the professor is woken up for his shift, he feels his head, and says 'How stupid is that barber? He's woken up the bald man instead of me.'[6]

The ambiguity in this case is one of identity. By laughing at the joke, we temporarily free up our assumptions about the identities of the two people who are being confused, though within the overall reassuring framework of a joke used for social communion. Identity mistakes are a common theme of comedy, especially in ancient Greek and Roman plays, and in Shakespeare's *Comedy of Errors* (which features two pairs of identical twins who are constantly mixed up). What makes this joke particularly effective, though, is that the professor is confused about *his own* identity rather than just someone else's. Though the immediate effect may be very temporary, this is a way of stretching our sense of meaning about the identities of people in a way that contributes to our total meaning resources and the flexibility of our weak neural channels. A minor balancing feedback loop replaces the reinforcing ones we may often experience when forming less flexible beliefs about people's identities.

The most obvious forms of ambiguity in jokes are puns, but I'd argue that the same points apply to slapstick. This usually involves acts of controlled violence or other minor taboo-breaking, but placed within a context of reassurance that tells us that the usual blame and harmfulness attached to such actions does not apply. Slapstick, as already mentioned, goes back to ancient Greece, but this time let's consider a modern example. When, in the classic cat-and-mouse cartoon *Tom and Jerry*, the cat Tom hits the mouse Jerry with a large hammer, so that Jerry at first appears to be flattened, we are then reassured as Jerry is miraculously resurrected, springing from 2D

---

6 'Classic gags discovered in ancient Roman joke book', *The Guardian* 13 March 2009.

into 3D again and running off. This interestingly, again, puts us in a larger context, through the breakdown of normal assumptions about the effects of a violent action in the meaning-world of the cartoon. We are led on some level to recognize that no matter how many cats kill mice, they are part of a larger system in which it is very unlikely that the mouse as a type will ever be eliminated (and where indeed the cat, counter-dependent on the mouse, would not want it to be). Even if we don't fully appreciate this larger system, our resources for thinking about the smaller system are slightly stretched, not only in relation to cats and mice, but in relation to all hunter-and-prey type relationships. We are thus primed a little bit more for receptivity and adaptation.

Satire similarly places things in an ambiguous frame, but this time in relation to authority, authoritative people, or unreflective social mores. Eighteenth-century Britain offered a golden age for satire (though again it goes back to ancient Greece and the work of Aristophanes). An example is Jonathan Swift's 'A Modest Proposal For Preventing the Children of Poor People From Being a Burthen to their Parents or Country, and For Making Them Beneficial to the Publick' of 1729.[7] Swift begins with a description of the problems of poverty, then comes in unexpectedly with a proposal to eat the excess children of the poor: 'A young healthy child well nursed, is, at a year old, a most delicious nourishing and wholesome food, whether stewed, roasted, baked, or boiled; and I make no doubt that it will equally serve in a fricassee, or a ragout.'[8] He contrasts this with a set of more ethical and practical proposals that he then rejects. Swift is here satirizing simplistic solutions to complex problems, but also utilitarian solutions where the end justifies the means but where the means are inhumane. He relies on us being horrified by his 'solution' but also seeing it in a wider context of reassurance that he does not seriously propose it. In this way we may be led to widen our understanding of narrow and simplistic 'solutions', to see them in a wider context of meaning that enables greater provisionality in relation to them, rather than engaging in the reinforcing feedback loops that would follow if we took such 'solutions' seriously.

7 Swift (1729).
8 Ibid.

When this integrative practice is hijacked for absolutizing ends, the ambiguity remains but is used, not to widen our awareness of meaning, but to reinforce a fixed belief in contrast to a rejected one. The Roman joke above could conceivably be used as an attack on supposedly absent-minded professors, and if we substituted an Irishman in a British context (or any other scorned nationality that has become the butt of jokes in another national context), it would become clearer that a particular group was being mocked, and by implication others were having their status reinforced by contrast. Slapstick can be interpreted as making violence more acceptable by dulling the edges of our disapproval to it, again with the ambiguity exploited by a dominant person or group who depicts others being physically attacked 'as a joke'. Satire can be genuinely mistaken for serious proposals, or it can extend the acceptability of absolutizing ways of thinking. 'Calculated ambiguity' appears to be a significant strategy of right-wing populists working in internet media or social media[9] – for instance, introducing antisemitic elements in a half-jokey way that makes it possible to claim that it was 'just a joke' if anyone objects, but that nevertheless adds to the influence and acceptability of absolutizing racist claims.

Though the basic dynamics of humour are unlikely to have changed very much since ancient times, the ways in which we access and spread it obviously have done so. Humour can be readily shared across the world by comedians using broadcast media, and by anyone using social media. This amplifies the potential power of humour both as an integrative practice and as an absolutizing one, making the potential balancing or reinforcing feedback loops bigger. Nevertheless, its healing power remains its most important characteristic. No matter how grimly the cycling conditions around us deteriorate, we can always laugh.

---

9   Gadinger & Simon (2019).

4.i. Reflection and Autobiography

> *Summary*
>
> Systematic reflection, whether or not written down in a journal, can be an integrative practice that widens our access to meaning and consideration of beliefs. In relation to one's life it is autobiography, a form of writing that can provide archetypal inspiration, or also reinforce dogmas. Journals and autobiographies go back to ancient times, with both ancient and modern examples showing their contrasting integrative or disintegrative effects.

Reflection can take many possible forms, but they all involve bringing greater attention to experience, probably also considering it in relation to different framing beliefs and assumptions. It can be relatively theoretical or relatively individual, but it always brings together theory with our individual experience, and is characterized by a certain discipline that distinguishes it from day-dreaming or rumination. One may adopt a structure (for instance, a set of framing questions) and think through that structure systematically, or think through a problem and how to approach it, or review a particularly important experience to work out its implications. When this reflection process focuses on individual experience over time, one can also call it *autobiographical* in a wide sense.

The process of reflection is integrative because it enables new meaning to emerge, and new beliefs to be provisionally considered, in relation to one's previous beliefs and experience. It may be particularly helpful to integrate it with kinaesthetic movement: the relationship between reflective thinking and solitary walking had been particularly important in my own experience. Nietzsche was also a great advocate of this, apparently saying 'Never trust a thought that didn't come by walking'.[1]

The results of such reflection may well then also be written down. This is often a helpful process, because it enables further reflectivity by incorporating meaning over time (if one reads back what one has written later), or from others. The form of the writing could, for instance, be that of a journal, an autobiographical writing, a fictionalized narrative, a poem, a diagram, or a more theoretical work such

---

1   I can't find a source for this, but it's a provocatively interesting thing to say even if Nietzsche didn't actually say it.

as this one (which is the result of a process of reflection over a long period). However, the value of reflection and autobiography is not limited to the written form. It could all be kept 'in one's head', perhaps through a system of memorization such as a 'memory palace',[2] or it could be told to others. One exercise that I've found highly beneficial myself is to tell my life story to a group of friends. However, written autobiography or memoir can also be used as a way of systematically reflecting on, and learning from, one's individual past.

Reflection as a practice is likely to go back at least as long as practices such as philosophical discourse, or of writing in poetry or prose. However, we do not have records of such reflections, unless they were recorded in a journal or memoir that was subsequently preserved. Religious practice provides the most common motive that gives a reflective context to experience, but the recording of public life, or of the novelties of travel, seem to have also provided a motive from an early stage. There are sections of the Pali Canon in which the Buddha is depicted as telling parts of his life story, such as the *Ariyapariyesana Sutta* or 'Discourse on the Noble Quest',[3] which tells the story from the Buddha's going forth to the first sermon. The autobiographical elements are obviously heavily filtered by the forms and language of later Buddhism, but they nevertheless give us some of the crucial symbolic story of the discovery of the Middle Way. Julius Caesar published a rather impersonal record of his conquests,[4] and the Roman Emperor Marcus Aurelius wrote a series of *Meditations* that have become a sourcebook of Stoicism.[5] There are several relatively early records of journeys in Chinese, Japanese, and Arabic traditions. However, perhaps the earliest reflective spiritual memoir is St Augustine's *Confessions* from 397–400 CE.[6]

These early examples already raise many of the issues accompanying journaling or memoir for public consumption. On an individual level, these practices enable balancing feedback loops of further meaning in relation to past experience. When made public, they can also model such practices, as well as potentially offering sources of archetypal inspiration (usually of a heroic type) for others. However, stories that originally had these functions can be

2   See Buzan (2003).
3   *Majjhima Nikaya 26*: Ñanamoli & Bodhi (1995) pp. 256ff.
4   Caesar (1869).
5   Marcus Aurelius (1964).
6   Augustine (1907).

wholly or partly hijacked by dogma – as in the presentation of the Buddha's life in the Pali Canon at times as part of a triumphalist enlightenment narrative. Public presentation can help to maintain repression: Caesar's apparently bluff and truthful narrative of the Gallic conquests reveals little about his motives or inner conflicts, contains some rather incredible claims, and was probably used for propaganda purposes to justify those conquests and boost his political standing.[7] St Augustine's *Confessions*, whilst providing a critical perspective on his earlier Manichaeism and his other youthful sins prior to his conversion to Christianity, is a conversion narrative that, unlike the Buddha's, incorporates a one-phase, not a two-phase self-critique. He does not go on to show us the limitations of his interpretation of Christianity, but rather leaves this as the triumphal end-point.

The Romantic movement of the late eighteenth and early nineteenth century provided a new impetus for the personal narrative centring on the individual, accompanied, as Iain McGilchrist points out, by a recursion to right-hemisphere functions. This gave a new power to the published individual memoir. William Wordsworth's verse-memoir, *The Prelude*,[8] is perhaps the most outstanding example of this in English. Here it is not only Wordsworth's reflections on past experiences themselves, but the role of 'Nature' that is crucial: the way that Wordsworth describes the landscapes in relation to which his experiences took shape is a constant mirror that shines greater light on them, and provides a new way of accessing his spiritual experience for the reader. Of course, Wordsworth is not without his idealizations and vanities, but it is clear that for him, autobiographical reflection in this form was the central practice enabling his character and judgement to integrate.

In modern times, we have a wide range of published autobiographical material that ranges from extensive and inspiring accounts of a lifelong learning process to the one-sided self-justificatory memoir. I will take just a couple of contrasting examples. Nelson Mandela's *Long Walk to Freedom* (1995)[9] includes details both of how its subject coped with long imprisonment, and negotiated a peaceful end to Apartheid in South Africa. It may have provided a vehicle of reflection for its author, but primarily provides a big source of

---

7   Hennige (1998).
8   Wordsworth (1995).
9   Mandela (1995).

inspiration for others in engaging with conflict. On the other hand, Prince Harry's recent ghost-written autobiography *Spare*[10] gives a long one-sided litany of sources of resentment between him and his royal family, with no attempt to contextualize this in awareness of others' perspectives. Autobiographies, like prehistoric fish, may be adaptive or maladaptive.

---

[10] Windsor (2023).

4.j. Travel and Foreign Languages

> *Summary*
>
> Travel and foreign language learning both go back to paleolithic times, and provide stimulus that is of huge value for the integration of meaning, providing resources for provisionality. The tradition of pilgrimage can also be of value by re-embodying the Middle Way, focusing on the process of travelling. Studies of bilingualism and translingualism show the value of learning foreign languages in increasing the flexibility of our use of models. These practices support globalization, but this is a process producing balancing as well as reinforcing loops, sometimes worth the trade-offs.

Travel as an integrative practice goes back to the paleolithic age. As already discussed in 3.a, many Native American and Australian Aboriginal groups wandered great distances and encountered members of many other groups. It was not transport technology that was most essential for this (they did not even have horses), but the conventions of a wider cultural area within which people felt a social duty to be hospitable to strangers. It is no coincidence that this unexpected degree of cosmopolitanism in certain early societies was accompanied by some of the key conditions for democracy (which I will return to in 4.l): for instance, the capacity to engage in discourse in spite of substantial differences, as a means of reaching consensual decisions. Encountering people of different culture beyond one's immediate group may well be a key contributor to enabling people to move from an interpersonal to an ideological stage of judgement, where such discourse becomes possible.

Apart from its likely role in stage development, the value of travel undertaken as an integrative practice at any stage should not be underestimated. There are a variety of reasons why 'travel broadens the mind' as the saying has it, consistent with the account of meaning and belief I have developed in Middle Way Philosophy. Unfamiliar environments bombard us with new meaning, creating a great many new weak neural links that we can then make subsequent use of to develop new adaptive beliefs. Encounters with new people outside our normal group also stimulate understanding of new potential beliefs that we may then enter into dialogue with – perhaps out of practical necessity in some cases. Sometimes the process of travel itself, even prior to the arrival in a new place, can also be integrative as recreation (see 4.k below): particularly if one adopts a slower mode such as walking or cycling.

The value of travel as an integrative practice is reflected in the long-standing recognition of pilgrimage in religious traditions. Pilgrimage can, of course, be an ascetic practice, merely consisting of inflicting hardship on oneself in the fixed belief that one will gain merit, or follow God's will, by doing so. However, the more one focuses on the journey as an experience, and the encounters that come with it, rather than the goal as an end in itself, the more likely it becomes that one will start to strike a balance between idealized belief in the goal and full recognition of the process and its difficulties. On a simple level, we can see this demonstrated in the ways that walking can help with depression.[1] Pilgrimage can also directly enact the Middle Way as a partially re-embodied metaphor, combining as it does the schemas of an ongoing path with the experience of balance:[2] we follow a path in a literal embodied sense, although the balance we need is still metaphorical as well as literal. That re-embodiment could well assist us in overcoming the effects of too much enculturated disembodied thinking.[3] We have to constantly pace ourselves, re-adjusting the balance between pursuing goals and addressing the conditions needed to pursue them, as we encounter new conditions of many kinds.

In medieval times, pilgrimage was a popular practice, which may sometimes have been ascetically motivated (such as for penance), or by the quest for a miraculous cure, but also often probably offered an implicit Middle Way discovered in the process of making the pilgrimage. In Christendom, pilgrimages to Jerusalem or Rome involved considerable distance for many, and thus danger and hardship. The same can be said of the Muslim *hajj*, or pilgrimage to Mecca, and of Buddhist pilgrimage to the sites associated with the Buddha in northern India. Just to make such a journey requires resilience and adaptability in the face of new and unexpected conditions, not just dogged determination or group conformity. Sometimes, also, there was awareness of the limitations imposed by too much goal-orientation in a pilgrimage:

1   Heesch et al. (2015).
2   See Ellis (2019) 3.b.
3   I.3.b.

There were Irish **peregrini** *(i.e. pilgrims) in the early Middle Ages who left their country and set out 'for the love of Christ' without a destination. They would get into boats without oars and let the wind, under the guidance of the Holy Spirit, blow them where it would.*[4]

This takes us to its obvious conclusion: the idea that pilgrimage is an integrative practice for us as individuals. If that is its focus, the destination actually does not matter, only that we are developing in the process of moving towards it. Of course, there may also be unnecessary perils in getting into a boat without oars! The same kind of motive is expressed by a modern Buddhist writer, Rijumati Wallis, who set out on a 'Pilgrimage to Anywhere':

All travel contains an inescapable element of uncertainty and confusion; whether it is the layout of a new town, a foreign language, alien food, or just the local vagaries of immigration and baggage handling. It is the way we respond to these confusions and uncertainties that reveals the different modes of travelling: are they a hassle to be endured so we can get on with enjoying ourselves, or are they a reminder of the assumptions we brought with us?.... As in Walt Whitman's iconic poem, the Open Road sang to me of freedom, a new beginning, opportunity, exploration, and transformation. I wanted to travel without a clear destination in mind, travel for a touch of mystery, danger and the unexpected.[5]

There is probably little need to elaborate, however, on all the ways that even for medieval pilgrims, let alone modern travellers, the majority of travel does very little to help create these balancing feedback loops, but rather sets up a whole lot of new reinforcing ones. For medieval pilgrims, those may have lain in metaphysical beliefs about the significance of the pilgrimage, its motive and destination – or at the other extreme a loss of any awareness of the sanctity marking out the journey as an integrative practice, when it was motivated only by group conformity. For modern travellers, on the other hand, mass tourism usually involves only very superficial encounters with a different host culture at one's destination, and instead depends strongly on the hedonic treadmill. We anticipate pleasure in the sensual experiences of a new location, but become so obsessed with these as conceptual tokens that they substitute for awareness of the experience itself – so the experience itself is mechanically reduced to an opportunity for snapshots and social media posts. Being able to say that we've been there becomes more

---

[4] Harpur (2016) p. 6.
[5] Wallis (2010) pp. 1-2.

important than actually being there, and we are thus unable to learn from being there. The negative environmental and socio-political impacts of tourism (such as air pollution, over-development, or undermining of local culture), rather than being to some extent an unavoidable trade-off for creating the conditions for an integrative experience, then instead become another reinforcing feedback loop tributary to the one we have created within ourselves.

However, in the modern context there are still ways of engaging with travel as an integrative practice that avoid many of these drawbacks with closed hedonic loops. Longer stays in a different culture that involve closer contact with the locals and fuller immersion in their environment are much more likely to be integrative: so working, volunteering, or studying abroad is often a much better way of travelling.

The closer one's immersion in another culture, however, the more the issue of foreign language use is likely to arise. Learning foreign languages not only widens our vocabulary, the range of people we can communicate with, and the material we can read, but more deeply, acquaints us with different categorizations and models, as well as different metaphorical developments of those models. This is obviously beneficial for the integration of meaning, creating a more diverse set of neural links and providing us with wider resources to draw on when developing possible beliefs.[6] To give a simple example, familiarity with the distinctions made in both French and German between two types of what in English is the same word, 'know' – *savoir* or *wissen* on the one hand and *connaitre* or *kennen* on the other – can provide an English-speaker with a prior familiarity with the ways in which, in practice, some 'knowing' is refers to belief and some to meaning. This can help to prevent confusions whereby we interpret meaning-knowledge (what philosophers call 'knowledge by acquaintance' – *kennen* or *connaitre*) as absolutized knowledge.

Studies of bilingualism have given further support to this point, by showing how bilingual individuals are able to resolve conflicting cues more easily.[7] Nor is it necessary to be fully bilingual, or even fluent, for some benefit in terms of integration of meaning to emerge from learning foreign languages: my own experience is more that of

---

6   II.6.e.
7   Bialystok (2009).

dabbling in a wide range of foreign languages, but this is one that I have found beneficial for providing an initial expectation of differing framing in different contexts. This also helps to break down any rigidity one might have about the boundaries of 'languages': such boundaries are provisional, written ineffectively on the ripples of a flowing river of *language*. The etymological sections of dictionaries chart the sometimes dried-out river beds.

It's very difficult to know when foreign language learning first began: but those paleolithic wanderers must have developed at least some degree of understanding of the language of the host communities they were passing through. In Aboriginal Australia, for instance, great language diversity does not seem to have been a barrier to travel, with the 'walkabout' having the status of a rite of passage for young men in Aboriginal culture. A study by Jill Vaughan on the complex fluidity in the use of different languages by members of indigenous communities in Arnhem Land, Northern Territory, shows ways that languages are closely associated with particular areas of land and particular social codes, but that individuals nevertheless switch between these quite readily, suiting the language not only to the auditor but also to the content of what they are communicating.[8] As the title of Vaughan's article indicates, 'translingualism' is 'ordinary', and has probably been ordinary for a very long time.

We should probably think of translingualism as an aspect of the cosmopolitanism, and thus of the complex ways of working with potential conflict, that we can find rather unexpectedly (at least from a modern Anglophone point of view) in all sorts of places through human cultures across space and time. In contrast, monolingualism provides one of the contributing conditions of meaning that enable absolutized socio-political ideologies to be sustained. That is not because it has always necessarily been a tool of power, but because the exertion of power – whether imperial, economic, or cultural – is much harder without it. The beliefs of imperialists need to be shared, bound together by absolutizations that rely in turn on particular shared sources of meaning, and many of these are likely to be codified in an influential language: Latin and English being the most salient examples during the last 2000 years in the West. If, like me, one comes from a monolingual background that has been

---

8   Vaughan (2020).

shaped by such past exertions of power, translingualism of any kind is a practice that can help to loosen absolute assumptions and thus the reinforcing feedback loops that accompany them.

The learning of foreign languages as an aspect of education goes back at least to the Roman Empire, where upper-class boys would be instructed in Greek, the culturally prestigious language. The learning of Latin was also crucial to nearly all Western education throughout the Middle Ages, just as the use of English has now largely become for much internationalized academic learning and research. In recent decades, language learning has adapted significantly from the highly formalized and explicit memorization of grammar and vocabulary to increasingly focus on the embodied situations in which language is used. This can not only make learning more effective for many students, but can expand the meaning it integrates by giving greater emotional weight to the language learnt. At the same time there are ways that too much of a functional focus can limit meaning by associating it too strongly with particular contexts: for example, we might learn enough modern Greek to order a meal in a taverna, but be unable to express our wonder at the ruins of Mycenae. A 'stretch' process in which we ground our learning of meaning strongly, but then also universalize it and thus keep applying it to new contexts, probably best identifies the Middle Way here. The formal learning of grammar in language education is thus not dead, nor ever likely to be so.

Both ease of worldwide travel and the increasing predominance of English, together with commercial and political interdependence and communication technologies, are aspects of 'globalization' as it has impacted the world from the beginnings of the European Empires in the fifteenth century through to the present. Whatever its one-sided detractors may claim, globalization is not solely in service to a bureaucratic, imperialist, or 'colonial' worldview, though these are aspects of how it has arisen. Rather there are both pros and cons to it. The breaking down of meaning barriers between cultures, and the development of greater international understanding and coordination, are of great positive value in creating the conditions for more integrated social judgements. Trade ('gentle commerce' as Steven Pinker calls it) has also been the realistic means by which much of this has been developed: you have to avoid the conditions of conflict with others when it is to your advantage to do so, because you trade with them.⁹ That globalization has also contributed to major reinforcing feedback loops in the exploitation of the environment is also clear, as is the immediate probability that this drawback will rapidly eclipse all the integrative advances that have been made. However, to deal with whatever immediate situation arises, cross-cultural understanding and international cooperation will be crucial, so we should never underestimate the value of travel and language learning as practices – even when they are only possible with costs and trade-offs.

9   Pinker (2011) p. 824.

4.k. Recreation and Green Environments

> *Summary*
>
> 'Recreation' is compensatory activity that has been made necessary by the over-specialization and alienation of work since the industrial revolution. In its context it is integrative and necessary, but also mild, temporary, and supportive of continuing reinforcing feedback of exploitation in the wider economic conditions. The constricting effects of artificial urban environments have also created a similar compensatory need for immersion in green environments, that can also be intensified into sublime experience. This is not 'nature', which is often projected, but it is contextually valuable.

'Recreation' is a much wider concept than any of the other practices I have considered so far, covering hobbies, games, and sports as well as some things that I have already mentioned such as the arts and bodywork. It is often equated with 'leisure' – that is, any activity that is not required to meet our immediate physiological needs or social duties, but is subtly different. The difference seems to lie in whether or not we have been 'de-created' before we need to be 're-created'. Leisure is sometimes bestowed on those whose life is largely composed of it, so do not need it to recover from anything – such as young children, the landed gentry of former ages, or the retired of today. Recreation, however, is a term that seems to have been first used in the fourteenth century for the rehabilitation of an invalid, but then, particularly after the industrial revolution, began to be used more generally for activities that people need to engage in to recover from their work. Obviously this tells us a lot about how work is viewed, as it embeds the assumption that work must be 'de-creative' or disintegrative rather than creative or integrative. Recreation, then, is very evidently an integrative practice in a way that leisure is not necessarily. It is a practice that provides compensation for particular circumstances of inner conflict.

Alienation from work, which is the root problem that 'recreation' compensates for, is a concept developed by Marx in his *Economic and Political Manuscripts* of 1844.[1] It explains this psychological effect in an entirely socio-economic and philosophical way (as is typical for Marx), by attributing it to the ownership of the means of production

1   Marx (1977).

in capitalism. Because the worker has no control over any of the process of production – the design, the materials, the manufacturing process, or the destination – Marx argued that he became alienated from the product, the act of production, the 'species-essence' and the relationships around the production. This theory influentially identifies the sources of alienation that people have felt and still feel around their jobs in industrial society, but also adopts unnecessary dogmatic elements in its claims about human 'essence', or that the non-ownership of the means of production is a necessary part of the conditions of alienation. It would be possible in some circumstances (albeit uncommon ones) for some workers to be happily and creatively engaged in very constrained conditions, to be 'bound in a nutshell, and yet the king of infinite space' as *Hamlet* has it, if those conditions did not create psychological conflict. Conversely, alienation from labour can still occur amongst the self-employed who own their means of production and control all their contracts. However, there is little doubt that contractual employment, especially in repetitive tasks that fail to engage human creativity, often does create psychological conflict: so 'alienation' is conflict and 'recreation' is integration.

'Recreation' can be seen as a part of the wider capitalist system, a way of patching up alienated workers sufficiently so that they can then go back to work for yet another round of soul-destruction. Whilst alienating work creates conflict through the use of power and absolutization, the conflicting beliefs it creates are liable to be long-term ones about the conditions (such as 'I have to do this job in order to pay my rent' versus 'I hate this job'). These conflicting beliefs can only be changed either through changing the conditions completely, or through very profound changes in habitual framing. Recreation does not 're-create' by addressing these, but only by addressing the immediate effects of the alienating work. Its over-specialization cramps the body (see 3.d on *deformation professionelle*), so recreation allows the body to be exercised in contrasting ways, for instance through sport. It prevents creative control, so recreations such as making model railways or playing video games are typically all about creating a compensatory virtual environment in which there can be creative control. Alienating work limits our relationships and often makes sure they are dominated by power-dynamics, so recreation enables us to associate with different people on the basis of shared interest – for example, in the local amateur dramatics society.

This is integrative practice, but of a kind that makes only a limited degree of temporary integration. Compared to meditation, the degree of temporary integration is in most cases likely to be limited in intensity, and to consist more in a sense of relief from the pressure of psychological conflict than in an experience of more intense integration. Even if we do experience some unification of energies that were previously caught up in proliferating reinforcing feedback, we are unlikely to be able to invest this much other than in coping with our work when we have to go back to it. If that is the use we need to make of our relatively integrated state at the moment, then it can be indeed an application of the Middle Way to use it in that way. However, in the larger pattern we can also see ourselves contributing to a bigger socio-economic feedback loop. Freeing ourselves from that is the bigger longer-term priority, which means loosening loops of exploitation of the kind discussed in 3.d.

One of the key impacts of alienating work, and of the capitalist system that has helped to produce it, is also that of the constricting artificial environment. Studies of the effects of immersion in green environments have shown the restorative value of such environments in reducing stress levels,[2] which has led to widespread talk of the value of being 'in nature'. In some senses, though, we are always 'in nature' (even a concrete underground car park, say, is part of the ecosystem), and in others, almost never, since nearly every environment has been greatly altered by human activity. The idea that by going for a walk on the open mountains of the Brecon Beacons in Wales near where I live, for instance, I will be having an experience of 'nature' is laughable – since this treeless green desert has been turned into its current monotonous state, very low in biodiversity, by sheep that humans have continued to put there. Nevertheless, if I had spent many successive days cooped up in an office gazing at a computer screen, walking on this green desert might well be good for me. Exposure to what (for lack of a better term, avoiding 'nature') we might call green environments – that is, outdoors and normally in the presence of plants and animals – is an extension of recreation, a compensation for over-specialized activity that is integrative due to the conditions created by that limitation. It is the conditions of

---

2   Berto (2014).

limitation (that is, the highly artificial environment such as office, supermarket, city street, apartment block) and their compensation that we should concentrate on here, rather than 'nature' as though it was anything other than a relatively normal and somewhat more balanced environment beyond this.

The artificial environment with its limitations is not an entirely new development in historical terms, but goes back to the development of cities in ancient times. The ancient Roman poet Juvenal, for instance, complained about excessive noise of traffic at night, congested streets made worse by the litters of the rich, and excrement or other rubbish being thrown out of windows.[3] It seems clear that ancient Rome was perhaps just as stressful an environment as many modern cities. It is thus not surprising that the contrasting appreciation of green environments goes back to a similar date, with Virgil's *Eclogues* developing an idealization of the countryside from an urban point of view that began in ancient Greece. This pastoral tradition in the arts continued after the Renaissance, and reached new heights in the Romantic movement from the eighteenth century. There, perhaps, the intensification of urban environments that was created by the industrial revolution was matched by an intensification of the pastoral idealization, which was no longer that of an orderly countryside so much as a wild 'nature'.

The ways in which this practice of balancing feedback loops can turn into reinforcing ones should thus also be evident: the Romantic commitment to 'nature' can become not just an experiential encounter with the wild phenomena beyond us, but a projection according to which ideological commitment to 'nature' will save us, neglecting all awareness of the ways we continually construct 'nature' for ourselves. While it remains an unprojected archetype, 'nature' can be an inspiring symbol, tapping into the God archetype to help us engage with new potential in our experience. When it is projected into metaphysics, however, 'nature' is every bit as infinitely manipulable as any other absolute: whether it's being used to bolster papal, scientific, political, artistic, or commercial authority.[4]

Positively, though, the recreational refreshment we may get from 'nature' may run through a whole spectrum, from a lunchtime stroll in the park to an extended immersion in the wilderness. What began

---

3   Juvenal, *Satires* 3: 238–77 (2001).
4   Ellis (2022) 6.a.

as, and most often remains, a limited compensation for the effects of absolutized boundaries in modern life, also does have its deep end: the sublime intensity of engagement with 'nature' that we may find in certain writers on the topic, and in the developing sphere of eco-spirituality. This quotation from Jane Goodall, about her experience whilst working with chimpanzees in a Central African forest, can stand for the kind of deep integrative experience and that can sometimes emerge from the intensity of modern compensation for the limitations of our self-created environment:

> Lost in awe at the beauty around me, I must have slipped into a state of heightened awareness.... I and the chimpanzees, the earth and trees and air, seemed to merge, to become one with the spirit power of life itself. The air was filled with a feathered symphony, the evensong of birds. I heard new frequencies in their music and also in singing insects' voices – notes so high and sweet I was amazed. Never had I been so intensely aware of the shape, the color of the individual leaves, the varied patterns of the veins that made each one unique. Scents were clear as well, easily identifiable: fermenting, overripe fruit; waterlogged earth; cold, wet bark; the damp odor of chimpanzee hair, and yes, my own too. And the aromatic scent of young, crushed leaves was almost overpowering....[5]

---

5   Goodall (2004).

4.l. Democracy

> *Summary*
>
> Democracy as an integrative practice is a form of decision-making that allows greater provisionality through the limitation of authority. This limitation of authority goes back to ancient, even paleolithic, times, but became increasingly effective with the gradual development of representative parliamentary democracy. It remains dependent on other features that developed from the Enlightenment – citizenship education, freedom of information, rule of law, and trust within an executive elite. This makes it dependent on other integrative practices, and vulnerable to authoritarian attack, appropriation, or corruption.

'Democracy' has become such a potent and abused word in the modern world, that at least two senses of it can be distinguished. It is most basically a system of government ('rule by the people') – though what exactly it means for the people to rule can vary a great deal. It is also, however, an ideal that can be understood as an archetypal symbol, as I discussed in *Archetypes in Religion and Beyond*.[1] It is as a system of government or of group decision-making that democracy can be an integrative practice. There, indeed, it seems to offer the best means yet devised for making collective decisions in a context of disagreement, whilst avoiding the absolutization of those disagreements and thus the development of conflict. It must, however, be broadly defined as a method, rather than pinned to a specific procedure, as has been the case with all the other integrative practices I have considered above. A brief consideration of its history as a practice can help to justify this point.

The crucial feature of democracy as a practice is the absence of any one authority dominating decision-making in a group or society. Although authorities are needed in practice in various spheres (for instance, generals in the army and judges in courts), they never have unconditional power in a democratic society, but are always subject to wider decision-making in the community. This seems to have been a feature of at least some paleolithic and mesolithic communities, as illustrated by the Wendat of North America, discussed by Graeber and Wengrow and already mentioned in 3.a. As then also discussed in 3.b., this feature does not seem to have

1   Ellis (2022) 6.g.

automatically disappeared with the rise of farming and early cities, but only to have been eventually disrupted by those who started using absolutization as a shortcut for social coordination to gain a short-term advantage (such as the rulers of Arslanstepe invading Sumeria). If Graeber and Wengrow are correct, then, 'democracy' in this wider sense is not something that suddenly appeared in ancient Athens, but a potential practice in social organization that can be found to varying degrees in a wide range of societies.

In the absence of any absolutized authority, some kind of decision-making system needs to be evolved. This could involve an oligarchic council of tribal elders, for instance, or decision-making by a class of male citizens, as occurred in ancient Athens. In ancient times this usually took the form of 'direct democracy' – namely, corporate decision-making within a selected group that in some sense represented the entire community. What is crucial at this point is not that everyone's viewpoint is directly represented (desirable though that may be), as that a variety of viewpoints are considered so as to enable provisionality in decision-making. Of course, though, the wider the variety of people represented, the wider the variety of viewpoints considered is likely to be. In most cases, democracies in this sense (whether they are formally called 'democracies' or not) do not represent everyone – perhaps women, younger people, foreigners, or subservient classes are not represented. These are also sometimes called 'republics' – early Rome, medieval Venice, and pre-Columbian Tlaxcala[2] all provide examples.

The transition from this to representative democracy, from late medieval times onwards, evolved initially from the English parliamentary system. What began as a meeting of barons to advise the king, after the Magna Carta was signed in 1215, soon added clerics and 'burgesses' – knights who represented towns, and who could be consulted on taxation. The powers of this parliament gradually expanded to include redress of grievances and law-making. The rights of the representatives to limit royal power having been established, they were then tested and extended in the English Civil War (1642–51), and Glorious Revolution (1689), when parliament deposed the king. The claims of justice and freedom thus gradually came to have institutional recognition in a modification of the existing structure of authority. Variants of this system were then widely

---

2   Graeber & Wengrow (2021) pp. 346–58.

adopted around the world in the form of more formalized written constitutions that balanced different powers (such as the US constitution of 1789).

Once the universal ideological values of justice and freedom were recognized in this way, though, it also created pressure for this system to be gradually modified to offer increasingly consistent representation to all individuals – including those without property, women, and young people at least over the age of 18. Discrimination of one group over another through slavery, racial segregation, or colonialism was also gradually and imperfectly reformed and reduced. Everyone in theory came to have a voice in Western democracy. In practice, those who actually wielded political power have until recently been wholly or largely limited to an elite, and the stability of democracy has been dependent on a degree of mutual trust within that elite.[3]

Along with the political philosophy of 'reason' arising during the Enlightenment came the evolution of state education for citizenship, of freedom of expression, and of the extending rule of law. Although modelling in terms of 'reason' tended to unhelpfully assume the absolute freewill of individuals in their judgements and the complete independence of the intellectual mind, at a sociopolitical level it did enable political systems to be successfully set up that allowed discourse between different groups and the peaceful resolution of conflict to occur. The system of political representation through elections and parliaments, despite many limitations, also generally provided at least enough channelling of political desires to enable the stability of democracy to continue.

There seems to be clear evidence that democracy helps to maintain peace. Although there are plenty of examples of supposed 'democracies' going to war with each other, when Russett and Oneal incrementalized and tested democratic systems in nations across the world from 1900, requiring a higher degree of maturity and stability in a democracy, they found a high level of correlation between mature, stable democracy and avoidance of war.[4] There are several reasons why this may the case: democracies have to pay attention to their citizens' suffering in war (or in opposition to it), they have an investment in the rule of law which they usually help

---

3   Pakulski (2012).
4   Russett & Oneal (2001).

to extend internationally, and they have an awareness and trust of democracy in other countries.[5] Although, in contexts from ancient Sumeria to Weimar Germany or modern Russia, democracies are vulnerable to attack and appropriation by unscrupulous absolutists, when they spread and work together, democracies develop an increasing defensive power and stability, along with a shared interest in joint defence when necessary. In many of the most recent conflicts, from the Second World War to the Russian invasion of Ukraine, democracies have been drawn into conflict, not with each other, but against absolutizing forces, either in clearly totalitarian states, or in anocracies like modern Russia that merely maintain some of the trappings of democracy but have not integrated it into their culture.

Democracy as an integrative practice, then, involves both participating in the democratic systems we may be fortunate enough to be part of, and supporting them against external threats. Participation means not only voting, but informing ourselves about what we are voting about, whether that is in a social organization, local government, national government, or international body. How else we navigate the complex balance between individual needs and political involvement is a question that I will need to go into in more detail later in this series.[6] Our capacity to deal with conflict and absolutization at an individual level is constantly interdependent with the ways in which conflict and absolutization are dealt with at a socio-political level, so we cannot simply ignore that level, but we can make clear decisions about it that take into account the amount of disruption it is capable of causing to our integrative practice at an individual level. This is, indeed, a corollary of treating democracy as a practice, rather than as a projected archetype. The practice does not itself resolve all our problems, nor is an institution that is to some extent 'democratic' necessarily integrative in its judgements. However, *in relation to other integrative practices*, democracy can help to create the conditions for overcoming conflict and absolutization. Every time we use it to create a new balancing feedback loop, by taking into account a new condition through a democratic mechanism, we are using democracy as a practice to help contribute to integrative effects.

5   Pinker (2011).
6   IX.

There have been some formalizations of this recognition in political thought. John Dewey, for instance, described democracy as 'a form of moral and spiritual association'[7] – a view that stands out in a philosophical landscape that is more likely to justify democracy in utilitarian terms. It may well be that democracy usually leads to greater happiness for all than its absence does, but that is not the chief reason for adopting it: rather, it forces us to listen to each other, and thereby adapt to a wider perspective on the human state. It doesn't have to include everyone at all times to fulfil its function, but it does have to provide routes for the inclusion of those who have been excluded so far. Without sufficient recognition of the rights of those out of power to allow this process of inclusion, the balancing feedback loop of democracy ceases.

Democracy as an integrative practice is constantly interactive with other systemic levels of human experience, and many of our failures to use it helpfully come from too narrow a focus on the socio-political level and neglect of the psychological, neural, and biological dimensions of political judgement. One of the major benefits of democracy is biological: the reduction in stress that comes from a degree of agency, and from some assurance that society will help you with your problems. Stress creates absolutizing judgements in which anxiety is looped with dogmatic assumptions,[8] so it is hardly surprising if those subjected to authoritarian government then make worse judgements that then maintain that kind of government, whilst those who feel more secure will maintain the democratic conditions for that security.

There are a variety of ways that democracy can be corrupted, but what they all have in common is the blocking of new understanding from new sources by the assertion of power, often using the concepts and vocabulary of democracy to cover its corruption. This can happen by direct corrupt intervention (e.g. ballot-stuffing), constitutional corruption (e.g. skewed voting systems, gerrymandering), corruption of legal processes that are intended to guarantee democratic processes, nepotism, cronyism, the domination of media outlets either by the state itself or by corporations with vested interests, and (recently) manipulation of social media algorithms. Many of these, for instance, came to the fore as significant concerns during

7   Dewey (1993) p. 59.
8   I.1.c (pp. 26–7).

the presidency of Donald Trump (2017–21), in a United States that previously seemed to have stabilized democracy. All such corruptions are not just disastrous for democracy because of their injustice, but also because they stop people becoming aware of new conditions, and thus more basically doom them to non-adaptation. At a time when climate change particularly demands adaptation, this is a pressing issue.

The prevalence and spread of such corruption at the time that I am writing makes it all the more important that those who value democracy as an integrative practice do their best to defend it. This means at all costs avoiding cynical abandonment of democratic institutions themselves, or of relatively reliable but imperfect sources of information. We practise democracy by remaining critically aware of the extent that democratic institutions can still prove resilient and continue to play their vital role even when there are some signs of corruption, not by abandoning democracy in despair at the first sign of trouble. Unfortunately, though, the successful practice of democracy depends on other integrative practices (particularly critical thinking, discussed below) to help a larger critical mass of people move onto the interindividual stage and avoid the characteristic 'flips' of absolute thinking constrained by ideology alone.

4.m. Psychotherapy

> *Summary*
>
> Psychotherapy is rooted in early guiding encounters between suffering people and those perceived as wiser, for healing or integration. As a modern practice it began with Freud's insights, though these were accompanied by causal speculations, and was developed as integrative practice particularly by Jung and Rogers. Although working within a number of tensions and prone to theoretical dogmas, it continues to offer valuable integrative practice.

Psychotherapy is an approach to integration that has developed out of healing arts, but that typically relies only on human-to-human communication between a person with more expertise or experience and one with less, rather than the use of drugs or surgery. There is some relationship to the role of shamans going back to paleolithic times, as the shaman (with or without mind-altering substances) led others into imaginative awareness of new possibilities that could either be seen as motivated by healing or as motivated by wider integration. The same could be said of any trusting relationship of guidance through the ages, when a person consulted a village wise woman, a guru, a philosopher, or an elder friend or relative. The immediate motive may well have been some form of suffering (an illness, a mania, depression, conflict in relationships), but the suffering was in many respects a prompt for addressing new conditions, that could only be addressed by developing new awareness. The boundary between 'healing' (returning from a state of suffering into a 'normal' state) and integration (overcoming conflicts in the long-term as part of the development of one's life) has thus never been as clear in this area as it might be in the healing of a broken bone or an infectious disease.

It is the formalization and professionalization of this process, distinct from medicine in general, that has given rise to the term 'psychotherapy' since the late nineteenth century. Some early nineteenth-century psychotherapists developed practices using hypnotic suggestion, but it is Sigmund Freud (1856–1939) who is widely recognized as the founding father of psychotherapy. It is Freud who first used techniques such as free association and dream interpretation to help his patients gain awareness of new meaning, and thereby gain awareness of the processes of projection that are continually reproducing conflict in our relationships. It is also Freud who began to recognize the relationships between conflict, repression, and fixed belief (psychosis or neurosis), which I have previously discussed in *Absolutization*.[1] Given the exclusivity of absolutized beliefs, conflict is continually created as they repress challenging beliefs, and the latter are then relegated to the 'unconscious', only to emerge unpredictably into potential awareness as we widen our awareness of meaning. Free association and dream interpretation are thus ways of developing contextual awareness of absolutization, and they have a close relationship with the arts in their integrative approach.

If Freud raised awareness of new kinds of integrative techniques that could produce balancing feedback loops in patients (raising awareness of conflict and helping them to re-frame it), many of his speculative theorizations then immediately followed this up with reinforcing feedback loops. One major focus of his theory was the attempt to identify causes of conflict in early childhood experience, an obsession that could lead to a great deal of confirmation bias, as he mined ambiguous dreams or slips of the tongue for confirmatory evidence of a particular childhood cause. As Popper later pointed out, Freud's work was unscientific in making no systematic attempt to counteract confirmation bias.[2] Freud's theory was also reductive in seeing all human desire as sexual desire, a framing that again could be constantly reinforced by one-sided interpretation of all possibly countervailing evidence.

However, Freud's work also stimulated the development of a wide range of schools of psychotherapy. Carl Jung was stimulated by Freud's understanding of conflict in mental illness and by some of his therapeutic approaches, but managed to avoid both his speculative obsession with causal explanations and his reductionism. For Jung, psychotherapy was never merely an application of a medical model to cure illness, but the outworking of a wider integrative vision that is most completely expressed in the first half of his *Red Book*.[3] This personal engagement with envisioned archetypal figures models an integrative process as it places those figures in a wider context of awareness, as well as sketching an explicit version of the

---

1   I.5.a.
2   Popper (1963) p. 37.
3   Jung (2009).

Middle Way itself.[4] The recognition of the archetype as a source of inspirational meaning in experience by Jung was thus another huge advance in psychotherapy, providing at least some of the core conditions for an engagement with religious symbolism that made use of its integrative power rather than getting caught up in absolute belief models.[5] All this has allowed many new balancing feedback loops through Jungian psychotherapy and the wider movement of Jungianism, as people have engaged with new aspects of their experience (with or without the aid of a therapist). The dangers of reinforcing feedback loops re-asserting themselves in the Jungian movement, however, are much more closely associated with the re-adoption of metaphysical beliefs that have often otherwise been discarded in modernity: Jung's own Gnosticism, which mars the end of his *Red Book*,[6] Platonic interpretations of the archetypes, and speculative beliefs about the 'collective unconscious',[7] none of which is functionally necessary to benefit from Jung's insights.

Fortunately, one of the ways that psychotherapy subsequently evolved was by questioning and discarding the role of interpretative theory and causal explanation in overcoming conflict. Carl Rogers's development of person-centred psychotherapy from the 1950s began with a recognition of the client's conflicts, which he called 'incongruence', and then used the therapeutic standpoint to help the person gain a wider awareness for resolving that conflict.[8] The honesty and humility of this approach is likely to guard against the therapy moving off into a new set of reinforcing feedback loops that are begun by the therapist's theoretical assumptions, but its limitation is also likely to lie in its conventionality. Rogers had a touching faith in every person's ability to work out their own healing process, but it may well be that they do not have the capacity to look outside the frame that is maintaining the conflict. Such approaches are also more limited by the medical model to bringing people back to an assumed 'normality', rather than stimulating them into the next phase of growth where the currently unthinkable might be encountered.

4   See Ellis (2020) for a much fuller account.
5   Ellis (2022) 1.g.
6   Ellis (2020) ch. 11.
7   Ellis (2022) 1.e&f.
8   Rogers (1979).

Of course, it is not possible here to survey all of the possibly thousands of types of psychotherapy that have been developed since Freud. In all of them, though, it can be fairly confidently asserted that the same types of tensions will be at work: the tension between the medical model and the integrative (or spiritual) model, the tension between helping the client to develop their own solutions and using a theoretical framework to generate stronger suggestions, and also the tension between examining beliefs themselves (as in cognitive behavioural therapy) and examining the wider emotional states that have sustained those beliefs. Any type of psychotherapy may potentially be used integratively, although some are better suited to overcoming specific problems ('neuroses' etc.) that block integration, rather than enabling integration more directly. Some forms of psychotherapy are heavily dependent on the use of a therapist, but others may be adapted as a basis of individual reflection. Studies of the effectiveness of psychotherapy in resolving specific problems have found evidence that it does work for this purpose, without much difference between types of psychotherapy.[9] That this study also found that psychotherapy works better in the long term than in the short term may also point towards the ways that healing is dependent on integration in the psychotherapeutic approach.

---

9   Seligman (1995).

4.n. Critical Thinking

> *Summary*
>
> Critical thinking is based on the wider application of certain philosophical and scientific skills to all judgement in every area of life, and began to be recognized as a separate practice during the twentieth century, mainly in an educational context. However, it still suffers from poor definition, particularly the perception that it is most centrally dependent on reasoning, when the range of skills it encompasses are much more those of contextualization. Putting it in the context of integrative practice could do much to clarify it.

Critical thinking is the last of the major integrative practices to gain anything like an explicit form. Indeed, I would suggest that it is still in the process of formation as I write today (in 2023). Although it has developed from philosophical enquiry, with its ancient roots discussed in 4.f above, critical thinking merely adopts some of the methods for examining beliefs that have been used in philosophy, and applies them much more widely. It is also congruent with scientific method as discussed in 3.g, but again, should not be appropriated as merely scientific, because it can be used by anyone in any context in relation to any material, regardless of its relationships to scientific method and results. However, the methods of statistical scrutiny used in the sciences are highly relevant to critical thinking.

The development of an explicit form of 'critical thinking', however, has arisen less in the context of either philosophy or of science than in that of education. John Dewey, writing in the 1920s, was one of the first thinkers to crystallize the need for a wider cultivation of critical skills as the basis of democracy[1] – particularly when democracy itself is an integrative practice (see 4.l). These critical skills, at the same time, are the ones required to move from the interpersonal to the ideological level of judgement, as they are typically applied in helping us recognize the limitations of beliefs that may be dominant in our background group, and justifying new beliefs that have at least an aspiration to universality. They are thus skills that are commonly required in higher education, and in the consistent practice of a profession in the modern context. However, the extent to which these skills were actually being taught in the context of existing

---

1   Dewey (1925) p. 437.

educational courses has often been patchy and haphazard. From the 1960s, then, an educational movement has gradually coalesced, mainly in the US, for the more systematic formulation and teaching of critical thinking skills.

If I say that this skill, as an integrative practice, is still in the process of formation, however, that is because it is still caught to a large extent within the assumptions of 'reason' and 'rationality'. As I argued in *Absolutization*, the idea that critical thinking is 'logical' thinking is an inflation of the significance of a deductive method that can only show the relationship between abstract concepts or hypotheses, to dominate an experiential practice that needs to be understood primarily in terms of the contextuality given to our justifications.[2] This inflation of logic is not aided by complex philosophical accounts of 'reason' or 'rationality' that try to recruit complex empirical features to the left-hemisphere dominated 'logical' process rather than recognizing that 'logic' was a conceptual substitution for the process of justification in experience in the first place.[3] At one level, at least, this point seems to have been implicitly obvious to the framers of critical thinking, who have often been educators rather than only philosophers, aware of the vast gulf between teaching students formal logic on the one hand and the actual practice by which they might bring critical awareness of their assumptions into judgement in a range of experienced cases on the other. However, this doesn't mean that they have been wholly able to shake off the rationalist legacy – perhaps because of a lack of clear alternatives as well as its general cultural entrenchment.

The consensual definitions of critical thinking that have emerged thus do show recognition that critical thinking is more than just logic, but nevertheless have failed to move on from the assumption that it is somehow dependent on logic. Nor have they found alternative ways of formulating it satisfactorily other than as a potentially contradictory hybrid. This is evident from the 'statement of expert consensus' on the subject arrived at by the American Philosophical Association:

> We understand critical thinking to be purposeful, self-regulatory judgment which results in interpretation, analysis, evaluation, and inference, as well as explanation of the evidential, conceptual, methodological, criteriological, or contextual considerations upon which that judgment is based.... The ideal critical

2   I.4.f.
3   I.4.b (pp. 102–4).

> *thinker is habitually inquisitive, well-informed, trustful of reason, open-minded, flexible, fair-minded in evaluation, honest in facing personal biases, prudent in making judgments, willing to reconsider, clear about issues, orderly in complex matters, diligent in seeking relevant information, reasonable in the selection of criteria, focused in inquiry, and persistent in seeking results which are as precise as the subject and the circumstances of inquiry permit.*[4]

What I find strange about such an obviously committee-derived definition is not that it is insufficiently comprehensive, but that it fails to identify any factor that unifies the traditional reasoning skills it starts with ('analysis, evaluation, and inference') with the more demanding aspects of embodied disposition that follow this up (such as being 'open-minded' and 'flexible'). There are many circumstances in which careful analysis and inference based on constricted premises is almost the opposite of open-minded, instead supporting the narrow left-hemisphere based rationalization so carefully documented by Iain McGilchrist in the model cases of patients with right-hemisphere lesions or schizophrenia. The father who put a coffin under the Christmas tree for his daughter (who was dying of cancer) was being entirely logical (she would indeed need it soon), but his premises lacked all awareness of living creatures and their emotions,[5] and his thinking (due to mental illness in this case) left these premises unquestioned. Rigorous analysis does not imply contextualization. On the other hand, those who are open-minded are also often lacking capacity for critical analysis, for instance failing to recognize the negative implications of the latest internet theory that they have credulously repeated. This 'definition' of critical thinking is quite extraordinary in its catch-all inability to focus on what is most important.

If, however, we begin to understand critical thinking in ways that sufficiently incorporate the implications of systems, psychology, and neuroscience, we can make much more sense of it. It is an integrative practice because critical *awareness* is its starting point: awareness that allows balancing feedback loops to challenge and replace reinforcing ones. In *The Five Principles* I offered an analysis of critical thinking in terms of six skills: reasoning, justification, recognizing contexts of argument, avoiding biases and fallacies, interpretation, and credibility assessment.[6] Of these, only the first

---

4   Facione (1990) p. 2.
5   McGilchrist (2021) p. 351.
6   II.6.g (pp. 255–66).

involves the logical connection of what are provisionally assumed to be the necessary relationships within a representation of the world that we are applying. We can only apply this linking skill appropriately, however, if we also apply all the other five skills, all of which are concerned with contextual awareness, and without which the logical connections we make have no justification. Justification, as I crucially argued, is not a matter of logical matching, but rather consists of a weight of associated experience that gives us confidence in a belief, and that may need to be weighed up against countervailing experience. Such weight of experience is our best bet as long as our judgement has not been hijacked by absolutization attempting to short-circuit it – so the avoidance of absolutization must thus also be an element of justification. All the other four skills involve aspects of awareness that help us to avoid absolutizing and thus gain better justified beliefs.

This analysis was developed from my personal experience of teaching critical thinking to many students, who benefited from it not because it improved their 'reasoning' skills, but because it prompted them to new levels of awareness of how they used any such skills, and because it honed and tested them in relation to a wide variety of issues.

We have now finally arrived at the opposite end of the spectrum from the place we began. The mere assertion of life as a reinforcing feedback loop develops increasingly complex forms both in biology and within human development and judgement. The most refined form of the reinforcing feedback loop is the dogmatic or metaphysical argument, in which logic is used to bind together the constant repeating of the same processes regardless of the context. The most refined form of balancing feedback loop thus responds to this by offering contextual awareness for such beliefs – not in order to simply negate them, but in order to reframe them in more adaptive ways. This refined, but nevertheless powerful, form of integrative practice is still not fully formed, but, regardless of that, needs to be applied urgently in whatever approximate forms we can manage, to prevent or at least limit the reinforcing feedback loops that are threatening to engulf our civilization. As a practice of recent development, though, it is interdependent with many of those that came before it, as well as on the whole structure of gradually more complex conditions charted through this book.

Conclusion

In this book I have tried to show a commonality of pattern between the most basic processes of life and the most complex judgements that we make as human beings. This is a commonality that runs through the 4 billion years of the history of life on earth, and consists in the operation of reinforcing and balancing feedback loops in the autocatalysed responses of organisms. Living organisms, from the beginning, were in danger of undermining their own conditions through multiplying loops of activity unresponsive to changes in the environment, but, also from the beginning, they found ways of adapting through electrical sensitivity. Now, on a massively more complex and accelerated scale, using human brains with trillions of neural links, each linked further into a complex globalized society, we are still facing the same basic challenge: how to adapt before we wipe ourselves out.

I have charted three different kinds of development that have interacted throughout that history to reproduce this pattern at ever greater levels of complexity: phylogenetic development through the evolution of species, ontogenetic development of individuals, and cultural development of human societies. Each of these types of development has revealed similar systemic patterns of development, in which reinforcing feedback loops in a stable environment continue for a while, but then gradually mount up frustrations until they reach a tipping point of change. Thus one species replaces another in an ecological niche, thus one psychological stage replaces the previous one in the development of a human individual, and thus a new cultural practice or form of organization continues until forced by new circumstances to adopt a new level of complexity. But no such succession is inevitable, and we do not know how long a system may get stuck.

I point out these patterns, not to advance any kind of new scientific theory or new historical account of human activity: primary researchers have done this kind of groundwork far more thoroughly than I could possibly do. Instead, my task has been to synthesize a systemic history that draws together threads from different disciplines for practical ends. Those practical ends are those of Middle Way Philosophy in general, and the ones advanced in this series of books as a whole: that is, to try to show why Middle Way practice is possible, practically necessary, and morally important. This book, by focusing on root conditions that make the Middle Way possible in the first place, is primarily about the first of these, though I hope it also makes a contribution towards illustrating its practical necessity and its moral importance from a certain point of view.

I began with some caveats about history, and I would like to repeat those caveats before I finish. History, even of this systemic kind, can only offer a plausible account of how conditions came to be as they are. Such accounts can easily be disputed and alternatives put forward. Why, for instance, should we want to account for historical phenomena in the way I have tried to do, in terms of reinforcing and balancing feedback loops? My answer is not strictly historical, but practical. I think it is more *helpful* to consider history in this way, than, for instance, to see it as a power-struggle between different forces. To see history as a pattern of reinforcing and balancing feedback loops shows both how our current problems and dilemmas are rooted in those loops, and offers a wider perspective in which they may not seem quite as overwhelming. Such a perspective determines nothing and guarantees nothing. It does not give you a source of authority as to what the infallible master really said, nor does it tell you what the solution always is because people got it perfectly right before. Nevertheless, you may find such a perspective helpful.

Organisms at different levels of complexity have faced the intractability of one species rigidly destroying the habitat of another, for instance, many times before. Similarly, humans have faced the limitations of their adaptation through psychological development many times: for instance, we may imagine, even back into medieval times, that educated ideological monks have sighed at the pigheadedness of interpersonal peasants who do not understand how a universalized perspective could help them. Similarly, cultural institutions and practices have been developed over and over again for human adaptation, and then been hijacked by those with a more limited perspective: religions have been fundamentalized, balanced political philosophies appropriated by demagogues, a carefully nurtured soil destroyed at a stroke by blasts of weedkiller. All this has happened before, many, many times. So our case, however overwhelming it seems now, is never unique. Whatever the momentum of reinforcing feedback loops that have developed before, balancing ones can always be found to start to address them.

Hope is always available, not because there is some cosmic source of it that we could possibly know about, but because our left hemispheres have it as a default setting, and return to it even after major disasters. However, it is important that we do not leave our left hemispheres unassisted by wider awareness, to hope blindly and naively. Instead, we need a consistent practical understanding of how to practise the Middle Way – how to tip the reinforcing feedback process in human judgement into a balancing one at any point.

For that practice, *The Five Principles* was the most central and seminal book in this series. If you have skipped it, then, I'd urge you to return to it as the heart of Middle Way Philosophy – that is, an account of how we can judge things differently in our lives, to actually make a difference. This book can only contextualize in particular ways, but closer reflection is needed on how to act. The further books in this series will be looking in more detail at how aspects of the background to the Middle Way connect with aspects of its practice in judgement: in judgements about meaning, in response to biases, in philosophical issues, in aesthetic issues, in ethical issues, and in political issues. They are more likely to attract those with a particular interest in any of these areas.

We live in a time when it is still possible, at least for those of us fortunate enough to have the practical security needed, to reflect in very general ways, to learn, and to prepare ourselves for the uncertainty to come. This kind of preparation is in my view far more important than any other. We do not know for what we are preparing ourselves, but there are strong indications that big disruptions are coming for human civilization. The best way we can prepare ourselves is to learn provisionality.

The Old and New Middle Way Philosophy Series

This book is the third of a planned series of at least nine books on Middle Way Philosophy, to be published by Equinox over the next few years. These books will together form a highly interconnected argument for the Middle Way as a practical philosophy. In the process they will synthesize various different sources of insight, and challenge various entrenched assumptions about human judgement, its justification and motivation. This series is in turn a substantial development, rewriting, and updating of an earlier series of four volumes.

While the new series is in the course of being written, I will need to refer the reader at a number of points to supporting and connecting arguments in both the old series and the new series. I suggest referring to the new series if possible, but using the old series if the required volume of the new series has not been published yet. To distinguish between them, I have used lower case Roman numerals (i, ii, iii, iv) in references to the books of the old series, and upper case Roman numerals (I, II, III etc.) to refer to the planned books of the new series. Both series are listed below.

Old series (Robert M. Ellis, 2012–15)

This has been published both as four separate volumes and as an omnibus edition by Lulu, Raleigh. This series is referred to using lower case Roman numerals, followed by section and chapter numbers.

i. *Middle Way Philosophy 1: The Path of Objectivity* (2012)
ii. *Middle Way Philosophy 2: The Integration of Desire* (2013)
iii. *Middle Way Philosophy 3: The Integration of Meaning* (2013)
iv. *Middle Way Philosophy 4: The Integration of Belief* (2015)

Middle Way Philosophy: Omnibus Edition (2015)

New series (Robert M. Ellis, 2022 onwards)

This is the third book of this planned series to be published by Equinox. This series is referred to using upper case Roman numerals. Obviously, references to books that have not yet been written (at the time of writing this book) must be approximate. I have sometimes given an indicative section number, but otherwise you will need to use the contents and index of the relevant book in the new series to locate a reference.

   I. *Absolutization: The Source of Dogma, Repression, and Conflict*
  II. *The Five Principles of Middle Way Philosophy: Living Experientially in a World of Uncertainty*
 III. *A Systemic History of the Middle Way: Its Biological, Psycho-developmental, and Cultural Conditions*
  IV. *Embodied Meaning and Integration: Overcoming the Abstracted Grip on Meaning in Theory and Practice*
   V. *Bias and the Middle Way: How to Stop Absolutizing our Conditioned Assumptions*
  VI. *The Practice of Agnosticism: Overcoming False Dualities across Human Thought*
 VII. *Mindful Beauty: Aesthetics as Gathering Attention*
VIII. *Middle Way Ethics: Stretching across the Gap between Relative and Absolute Values*
  IX. *The Middle Way Manifesto: Combining Radical Change with Political Effectiveness*

Bibliography

Adena, Maja, Enikolopov, Ruben, Petrova, Maria, Santarosa, Veronica, & Zhuravskaya, Ekaterina (2015) 'Radio and the rise of the Nazis in prewar Germany', WZB Discussion Paper, No. SP II 2013-310r, Wissenschaftszentrum Berlin für Sozialforschung, Berlin. https://doi.org/10.2139/ssrn.2603589

Ainsworth, Mary, Blehar, Mary, Water, Everett, & Wall, Sally (2015) *Patterns of Attachment: A Psychological Study of the Strange Situation*. Psychology Press, New York.

Allen, David & Howell, James (eds.) (2020) *Groupthink in Science: Greed, Pathological Altruism, Ideology, Competition, and Culture*. Springer, Cham.

Augustine, St, of Hippo, trans. Pusey (1907) *Confessions*. Dent, London.

Barandiaran, Xabier & Moreno, Alvaro (2008) 'Adaptivity: from metabolism to behavior', *Adaptive Behavior* 16:5, pp. 325–44. https://doi.org/10.1177/1059712308093868

Basavaraddi, Ishwar (2015) 'Yoga: its origin, history, and development' *Ministry of External Affairs of the Government of India*. http://www.redtwigyoga.com/uploads/1/2/1/9/12195443/yoga__its_origin_history_and_development.pdf

Beck, Don & Cowan, Christopher (1996) *Spiral Dynamics: Mastering Values, Leadership and Change*. Blackwell, Oxford.

Beckwith, Christopher (2015) *Greek Buddha: Pyrrho's Encounters with Early Buddhism in Central Asia*. Princeton University Press, Princeton.

Benesch, Oleg (2016) 'Reconsidering Zen, Samurai, and the Martial Arts', *The Asia-Pacific Journal* 14:17, p. 7.

Bernstein, Harris & Bernstein, Carol (2010) 'Evolutionary origin of recombination during meiosis', *Bioscience* 60:7, pp. 498–505. https://doi.org/10.1525/bio.2010.60.7.5

Berto, Rita (2014) 'The role of nature in coping with psycho-physiological stress: a literature review on restorativeness', *Behavioral Sciences* 4:4, pp. 349–409. https://doi.org/10.3390/bs4040394

Bialystok, Ellen (2009) 'Bilingualism: the good, the bad, and the indifferent' *Bilingualism: Language and Cognition* 12:1, pp. 3–11. https://doi.org/10.1017/S1366728908003477

Blackburn, Simon (1994) *The Oxford Dictionary of Philosophy*. Oxford University Press, Oxford.

Bowlby, John (1977) 'The making and breaking of affectional bonds: I: aetiology and psychopathology in the light of attachment theory', *British Journal of Psychiatry* 130, pp. 201–10. https://doi.org/10.1192/bjp.130.3.201

Brenner, Eric & 5 others (2006) 'Plant neurobiology: an integrated view of plant signalling', *Trends in Plant Science* 11:8, pp. 413–19. https://doi.org/10.1016/j.tplants.2006.06.009

Brodbeck, Frank, Hapla, Frantisek, & Mitlöhner, Ralph (2003) 'Traditional forest gardens as "safety net" for rural households in Central Sulawesi, Indonesia', Paper presented at The International Conference on Rural Livelihoods, Forests and Biodiversity 2003. https://www.researchgate.net/profile/Ralph-Mitloehner-2/publication/260386938_Traditional_forest_gardens_as_safety_net_for_rural_households_in_Central_Sulawesi_Indonesia

Bruner, Emiliano, Battaglia-Mayer, Alexandra, & Kaminiti, Roberto (2022) 'The parietal lobe evolution and the emergence of material culture in the human genus', *Brain Structure and Function*. https://doi.org/10.1007/s00429-022-02487-w

Bryant, Edwin (2016) 'Samadhi in the Yoga Sutras' from Eifring, ed., *Asian Traditions of Meditation*. University of Hawaii Press, Honolulu.

Bryden, M. (1963) 'Ear preference in auditory perception' *Journal of Experimental Psychology* 65:1, pp. 103–5. https://doi.org/10.1037/h0042773

Budd, Graham (2015) 'Early animal evolution and the origins of nervous systems', *Philosophical Transactions of the Royal Society B* 370:20150037. https://doi.org/10.1098/rstb.2015.0037

Burton, David (2001) 'Is Madhyamaka Buddhism really the Middle Way? Emptiness and the problem of nihilism', *Western Buddhist Review* 3, pp. 177–95. https://doi.org/10.1080/14639940108573749

Buzan, Tony (2003) *Use your Memory*. BBC Worldwide, London.

Caesar, Julius, trans. McDevitt & Bohn (1869) *Gallic War*. Harper, New York, and Perseus Project: https://www.perseus.tufts.edu/hopper/text?doc=Caes.+Gal.+toc

Capra, Fritjof & Luisi, Pier Luigi (2014) *The Systems View of Life: A Unifying Vision*. Cambridge University Press, Cambridge.

Christian, David (2005) *Maps of Time: An Introduction to Big History*. University of California Press, Berkeley.

Cinelli, Matteo & 7 others (2022) 'Conspiracy theories and social media platforms', *Current Opinion in Psychology* 47, 101407. https://doi.org/10.1016/j.copsyc.2022.101407

Cirotteau, Thomas, Kerner, Jennifer, & Pincas, Eric, trans. Hurd (2021) *Lady Sapiens: Breaking Stereotypes about Prehistoric Women*. Hero, London.

Conklin, Beth (2001) *Consuming Grief: Compassionate Cannibalism in an Amazonian Society*. University of Texas Press, Austin.

Connell, Joseph (1961) 'The influence of interspecific competition and other factors on the distribution of the barnacle *Chthamalus stellatus*', *Ecology* 42, pp. 710–23. https://doi.org/10.2307/1933500

Cooper, Robin (1996) *The Evolving Mind: Buddhism, Biology and Consciousness*. Windhorse, Birmingham.

Cowen, Tyler & Southwood, Ben (2019) 'Is the rate of scientific progress slowing down?', *GMU Working Papers in Economics* 21:13. https://dx.doi.org/10.2139/ssrn.3822691

Danziger, Shai, Levav, Jonathan, & Avnaim-Pesso, Liora (2011) 'Extraneous factors in judicial decisions', *Proceedings of the National Academy of Sciences* 108:17, pp. 6889–92. https://doi.org/10.1073/pnas.1018033108

Dawkins, Richard (1996) *Climbing Mount Improbable*. Viking, London.

De Geus, Marius (2014) 'Peter Kropotkin's anarchist vision of organization', *Ephemera: Theory and Politics in Organization* 14:4, pp. 853–71.

DeLellis, Pietro & 6 others (2014) 'Collective behaviour across animal species', *Scientific Reports* 4: 3723. https://doi.org/10.1038/srep03723

Descartes, René, trans. Sutcliffe (1968) *Discourse on Method and the Meditations*. Penguin, London.

De Villiers, Jill & De Villiers, Peter (1978) *Language Acquisition*. Harvard University Press, Cambridge Mass.

Dewey, John (1925) *Experience and Nature*. Open Court Publishing, Chicago.

Dewey, John, ed. Morris & Shapiro (1993) *The Political Writings*. Hackett, Indianapolis.

Diamond, Jared (1991) *The Rise and Fall of the Third Chimpanzee*. Radius, London.

Diamond, Jared (2005) *Collapse: How Societies Choose to Fail or Survive*. Penguin, London.

Diamond, Jared (2012) *The World until Yesterday: What Can we Learn from Traditional Societies?* Penguin, London.

Dickersin, K. (1990) 'The existence of publication bias and risk factors for its occurrence', *Journal of the American Medical Association* 263:10, pp. 1385–9. https://doi.org/10.1001/jama.1990.03440100097014

Di Paolo, Ezequiel (2005) 'Autopoiesis, adaptivity, teleology, agency', *Phenomenology and the Cognitive Sciences* 4, pp. 429–52. https://doi.org/10.1007/s11097-005-9002-y

Dumont, Louis (1985) 'A modified view of our origins: the Christian beginnings of modern individualism', ch. 5 from Carrithers, Collins, & Lukes, eds., *The Category of the Person: Anthropology, Philosophy, History*. Cambridge University Press, Cambridge.

Dunbar, Robin (1992) 'Neocortex size as a constraint on group size in primates', *Journal of Human Evolution* 22:6, pp. 469–93. https://doi.org/10.1016/0047-2484(92)90081-J

Dunbar, Robin (2004) *The Human Story: A New History of Mankind's Evolution*. Faber, London.

Dutt, Nripendra Kumar (1970) *The Aryanisation of India*. K.L. Mukhopadhyay, Calcutta.

Eifring, Halvor (2016) 'What is meditation?' and 'Types of meditation' from Eifring, ed., *Asian Traditions of Meditation*. University of Hawaii Press, Honolulu.

Ellis, Robert M. (2001) 'A Buddhist Theory of Moral Objectivity', PhD thesis, Lancaster University. Also published as *A Theory of Moral Objectivity* (2011) Lulu, Raleigh.

Ellis, Robert M. (2011) 'Taking the "Meta" out of Physics: a response to Graham Smetham's "The Matter of Mindnature"', *Journal of Consciousness Exploration & Research* 2:8, pp. 1086–1113.

Ellis, Robert M. (2017) 'Reason is not objectivity: a response to Julian Baggini's narrowly rational criteria for objectivity'. https://www.researchgate.net/publication/313696637_Reason_is_not_objectivity_A_response_to_Julian_Baggini%27s_narrowly_rational_criteria_for_objectivity

Ellis, Robert M. (2018) *The Christian Middle Way: The Case against Christian Belief but for Christian Faith*. Christian Alternative, Winchester.

Ellis, Robert M. (2019) *The Buddha's Middle Way: Experiential Judgement in his Life and Teaching*. Equinox, Sheffield.
Ellis, Robert M. (2020) *Red Book, Middle Way: How Jung Parallels the Buddha's Method for Human Integration*. Equinox, Sheffield.
Ellis, Robert M. (2022) *Archetypes in Religion and Beyond: A Practical Theory of Human Integration and Inspiration*. Equinox, Sheffield.
Erikson, Erik (1982) *The Life Cycle Completed*. Norton, New York.
Facione, Peter (1990) 'Critical thinking: a statement of expert consensus for purposes of educational assessment and instruction', American Philosophical Association.
Farazmand, Ali (2009) 'Bureaucracy, administration, and politics: an introduction' from Farazmand, ed., *Bureaucracy and Administration*. CRC Press, Boca Raton Florida.
Ferreiro, Emilia (1985) 'Literacy development' from Olson, Torrance, & Hildyard, eds., *Literacy, Language and Learning*. Cambridge University Press, Cambridge.
Finnerty, John (2005) 'Did internal transport, rather than directed locomotion, favor the evolution of bilateral symmetry in animals?', *Bioessays* 27:11, pp. 1174-80. https://doi.org/10.1002/bies.20299
Fischer, Richard, Alexander, Michael, Gabriel, Cynthia, Gould, Elaine, & Milione, Janet (1991) 'Reversed lateralization of cognitive functions in right handers', *Brain* 114, pp. 245-61.
Fondacaro, Rocco & Higgins, E. Tory (1985) 'Cognitive consequences of communication mode' from Olson, Torrance, & Hildyard, eds., *Literacy, Language and Learning*. Cambridge University Press, Cambridge.
Franks, Andrew & Scherr, Kyle (2015) 'Using moral foundations to predict voting behavior: regression models from the 2012 U.S. Presidential Election', *Analyses of Social Issues and Public Policy* 15:1, pp. 213-32. https://doi.org/10.1111/asap.12074
Freund, Alexandra & Ritter, Johannes (2009) 'Midlife crisis: a debate', *Gerontology* 55, pp. 582-591. https://doi.org/10.1159/000227322
Fujita, Frank & Diener, Ed (2005) 'Life satisfaction set point: stability and change', *Journal of Personality and Social Psychology* 88:1, pp. 158-64. https://doi.org/10.1037/0022-3514.88.1.158
Gadinger, Frank & Simon, Elena (2019) 'Calculated ambiguity, mobilized fear and the performance of ordinariness: right-wing populist narrative strategies in election campaigning and governing practices', *Zeitschrift für Politikwissenschaft* 29, pp. 23-52. https://doi.org/10.1007/s41358-019-00176-5
Gajduschek, Gyorgy (2003) 'Bureaucracy: Is it efficient? Is it not? Is that the question? Uncertainty reduction: an ignored element of bureaucratic rationality', *Administration and Society* 34:6, pp. 700-23. https://doi.org/10.1177/0095399702239171
Garstang, Walter (1922) 'The theory of recapitulation: a critical re-statement of the biogenetic law', *Zoological Journal of the Linnaean Society* 35:232, pp. 81-101. https://doi.org/10.1111/j.1096-3642.1922.tb00464.x
Gibson, Eleanor & Pick, Anne (2000) *An Ecological Approach to Perceptual Learning and Development*. Oxford University Press, Oxford.

Godfrey-Smith, Peter (2020) *Metazoa: Animals, Mind and the Birth of Consciousness*. William Collins, London.

Goel, Vinod & five others (2007) 'Hemispheric specialization in human prefrontal cortex for resolving certain and uncertain inferences', *Cerebral Cortex* 17:10, pp. 2245-50. https://doi.org/10.1093/cercor/bhl132

Gombrich, Richard (1988) *Theravada Buddhism: A Social History from Ancient Benares to Modern Colombo*. Routledge, London.

Goodall, Jane (2004) *Reason for Hope: A Spiritual Journey*. Grand Central Publishing, New York.

Graeber, David & Wengrow, David (2021) *The Dawn of Everything: A New History of Humanity*. Penguin, London.

Graves, Clare W. (1970) 'Levels of existence: an open system theory of values', *Journal of Humanistic Psychology* 10:2, pp. 131-55. https://doi.org/10.1177/002216787001000205

Griffiths, Mark (1995) 'Technological addictions', *Clinical Psychology Forum*. https://www.researchgate.net/publication/284665745_Technological_addictions (accessed 2022).

Gurevich, Andrew (2013) 'Forgotten wisdom of the Chauvet Cave: the sacred feminine and the (re)birth of culture', *Popular Archaeology*.

Haidt, Jonathan (2012) *The Righteous Mind: Why Good People are Divided by Politics and Religion*. Penguin, London.

Hardie, Jim (2017) 'Life cycles and polyphenism' from Van Emden & Harrington, eds., *Aphids as Crop Pests* (2nd Edition). CABI, Wallingford.

Harpur, James (2016) *The Pilgrim Journey: A History of Pilgrimage in the Western World*. Lion Hudson, Oxford.

Hashim, Rosnani, Rufai, Saheed, & Nor, Mohd (2011) 'Traditional Islamic education in Asia and Africa: a comparative study of Malaysia's *Pondok*, Indonesia's *Pesantren* and Nigeria's traditional *Madrasah*', *World Journal of Islamic History and Civilization* 1:2, pp. 94-107.

Hauser, Marc, Chomsky, Noam, & Fitch, Tecumseh (2002) 'The faculty of language: what is it, who has it, and how did it evolve?', *Science* 298, pp. 1569-79. https://doi.org/10.1126/science.298.5598.1569

Heap, E.G. (1967) 'Some notes on cannibalism among Queensland Aborigines 1824-1900', *Queensland Heritage* 1:7, pp. 25-9.

Heesch, Kristiann & 4 others (2015) 'Physical activity, walking, and quality of life in women with depressive symptoms', *American Journal of Preventive Medicine* 48:3, pp. 281-91. https://doi.org/10.1016/j.amepre.2014.09.030

Hennigar, Mallory (2022) 'Building Ambedkar's India: Nagaloka Centre as a microcosm of *Prabuddha Bharat*', *South Asian History and Culture*. https://doi.org/10.1080/19472498.2022.2159121

Hennige, David (1998) 'He came, he saw, we counted: the historiography and demography of Caesar's Gallic numbers', *Annales de Démographie Historique* 1998:1, pp. 215-42. https://doi.org/10.3406/adh.1998.2162

Henning, Stanley (2018) *Henning's Scholarly Works on Chinese Combative Traditions*. Via Media Publishing, Santa Fe.

Hibbing, Michael, Fuqua, Clay, Parsek, Matthew, & Peterson, S. Brook (2010) 'Bacterial competition: surviving and thriving in the microbial jungle',

National Review of Microbiology 8:1, pp. 15–25. https://doi.org/10.1038/nrmicro2259

Houellebecq, Michel, trans. Bowd (2005) *The Possibility of an Island*. Phoenix, London.

Hoverstadt, Patrick (2022) *The Grammar of Systems: From Order to Chaos and Back*. Scio Publications.

Hume, David (1975) *An Enquiry Concerning Human Understanding and Concerning the Principles of Morals*. Oxford University Press, Oxford.

Itzkin, Elissa (1978) 'Bentham's *Chrestomathia*: legacy to English education', *Journal of the History of Ideas* 39:2, pp. 303–16. https://doi.org/10.2307/2708782

Jardri, Renaud & 5 others (2012) 'Assessing fetal response to maternal speech using a noninvasive functional brain imaging technique', *International Journal of Developmental Neuroscience* 30, pp. 159–61. https://doi.org/10.1016/j.ijdevneu.2011.11.002

Jaspers, Karl, trans. Bullock (1953) *The Origin and Goal of History*. Routledge, London.

Jekely, Gaspar, Keijzer, Fred, & Godfrey-Smith, Peter (2015) 'An option space for early neural evolution', *Philosophical Transactions of the Royal Society B* 370: 20150181. https://doi.org/10.1098/rstb.2015.0181

Jekely, Gaspar, Paps, Jordi, & Nielsen, Claus (2015) 'The phylogenetic position of ctenophores and the origin(s) of nervous systems', *Evodevo* 6:1, pp. 1–9. https://doi.org/10.1186/2041-9139-6-1

Johnson, Mark (2007) *The Meaning of the Body: Aesthetics of Human Understanding*. University of Chicago Press, Chicago.

Joy, Melanie (2003) 'Psychic numbing and meat consumption: the psychology of carnism', Saybrook University ProQuest Dissertations Publishing, 2003:3073692.

Jung, Carl, ed. Shamdasani (2009) *The Red Book: Liber Novus*. Norton, New York.

Juvenal, trans. Kline (2001) *Satires*. Poetry in Translation: https://www.poetryintranslation.com.

Kahneman, Daniel (2011) *Thinking Fast and Slow*. Penguin, London.

Kapparis, Konstantinos, Arnaoutoglou, Ilias, & Spatharas, Dimos (2017) 'Clowning and slapstick in Aristophanes' from Macdowell et al., eds., *Studies in Greek Law, Oratory and Comedy*. Routledge, London.

Keeling, Patrick & 7 others (2005) 'The tree of eukaryotes', *Trends in Ecology and Evolution* 20:12, pp. 670–676. https://doi.org/10.1016/j.tree.2005.09.005

Kegan, Robert (1982) *The Evolving Self: Problem and Process in Human Development*. Harvard University Press, Cambridge Mass.

Kegan, Robert (1994) *In over our Heads: The Mental Demands of Modern Life*. Harvard University Press, Cambridge Mass.

Keijzer, Fred & Arnellos, Argyris (2017) 'The animal sensorimotor organization: a challenge for the environmental complexity thesis', *Biology and Philosophy* 32, pp. 421–41. https://doi.org/10.1007/s10539-017-9565-3

Keith, Arthur (1992) *The Sanskrit Drama in its Origin, Development, Theory and Practice*. Motilal Banarsidass, Delhi.

Killen, Shaun, Marras, Stefano, Nadler, Lauren, & Domenici, Paolo (2017) 'The role of physiological traits in assortment among and within fish shoals',

Philosophical Transactions of the Royal Society B 372, 20160233. https://doi.org/10.1098/rstb.2016.0233

Kinsbourne, Marcel & Bemporad, Brenda (1984) 'Lateralization of emotion: a model and the evidence' from Fox & Davidson, eds., *The Psychobiology of Affective Development*. Psychology Press, London.

Klikauer, Thomas (2015) 'What is managerialism?', *Critical Sociology* 41:7-8, pp. 1103–19. https://doi.org/10.1177/0896920513501351

Klimek, Peter, Hanel, Rudolph, & Thurner, Stefan (2009) 'Parkinson's Law quantified: three investigations on bureaucratic inefficiency', *Journal of Statistical Mechanics* 2009-P03008. https://doi.org/10.1088/1742-5468/2009/03/P03008

Ko, Kwang Hyun (2017) 'A brief history of imperial examination and its influences', *Society* 54: 272–8. https://doi.org/10.1007/s12115-017-0134-9

Kohlberg, Lawrence & Hersh, Richard (1977) 'Moral development: a review of the theory', *Theory into Practice* 16:2, pp. 53–9. https://doi.org/10.1080/00405847709542675

König-Weichhardt, Michael (Master Ziji) (2017) 'The father of tai chi and a mysterious immortal Zhang Senfeng' (blog post). https://internalwudangmartialarts.com/2017/04/17/the-father-of-tai-chi-and-a-mysterious-immortal-zhang-sanfeng/ (accessed 2023).

Konner, Melvin (2017) 'Hunter-gatherer infancy and childhood' from Hewlett and Lamb, eds., *Hunter-Gatherer Childhoods*. Routledge, London.

Kropotkin, Peter (1989) *Mutual Aid: A Factor of Evolution*. Black Rose Books, Montreal.

Kuhn, Steven & Stiner, Mary (2006) 'What's a mother to do? the division of labor among Neandertals and modern humans in Eurasia', *Current Anthropology* 47:6, pp. 953–80. https://doi.org/10.1086/507197

Kuhn, Thomas (1996: 3rd Edition) *The Structure of Scientific Revolutions*. Chicago University Press, Chicago.

Kuzminski, Adrian (2008) *Pyrrhonism: How the Ancient Greeks Reinvented Buddhism*. Rowman & Littlefield, Lanham.

Lacalli, Thurston (1994) 'Apical organs, epithelial domains, and the origin of the chordate central nervous system', *American Zoologist* 34, pp. 533–41. https://doi.org/10.1093/icb/34.4.533

Lakoff, George (2002) *Moral Politics: How Liberals and Conservatives Think*. University of Chicago Press, Chicago.

Larkin, Philip (1974) *High Windows*. Faber, London.

Laszlo, Ervin, Artigiani, Robert, Combs, Allan, & Csáanyi, Vilmos (1996) *Changing Visions: Human Cognitive Maps: Past, Present and Future*. Praeger, Westport Connecticut.

Lawson, Russell (2004) *Science in the Ancient World: An Encyclopedia*. ABC-CLIO, Santa Barbara California.

Lévi-Strauss, Claude, trans. Mehlman and Leavitt (2021) *Wild Thought (La Pensée Sauvage)*. University of Chicago Press, Chicago.

Lewis, Bernard (2002) *What Went Wrong? Western Impact and Middle Eastern Response*. Orion Books, London.

Liebermann, Philip (2006) 'The recent origin of human speech', *The Journal of the Acoustical Society of America* 119, p. 3441. https://doi.org/10.1121/1.4786937

Ling, Trevor (1979) *Buddhism, Imperialism and War*. Unwin, London.

Lodé, Thierry (2012) 'Sex and the origin of genetic exchanges', *Trends in Evolutionary Biology* 4:e1, pp. 1–5. https://doi.org/10.4081/eb.2012.e1

MacIntyre, Alasdair (1981) *After Virtue*. Duckworth, London.

Mallatt, Jon & Chen, Jun-Yuan (2003) 'Fossil sister group of craniates: predicted and found', *Journal of Morphology* 258:1, pp. 1–31. https://doi.org/10.1002/jmor.10081

Mandela, Nelson (1995) *Long Walk to Freedom*. Abacus, London.

Marcus Aurelius, trans. Staniforth (1964) *Meditations*. Penguin, London.

Markstrom, Carol, Berman, Rachel, Sabino, Vicki, & Turner, Bonnie (1998) 'The ego virtue of fidelity as a psychosocial rite of passage in the transition from adolescence to adulthood', *Child and Youth Care Forum* 27:5, pp. 337–54. https://doi.org/10.1007/BF02589260

Martineau, Jonathan (2015) 'Making sense of the history of clock-time: reflections on Glennie and Thrift's *Shaping the Day*', *Time and Society* 26:3, pp. 1–16. https://doi.org/10.1177/0961463X15577281

Marx, Karl, trans. Dutt (1977) *Economic and Philosophic Manuscripts of 1844*. Progress Publishers, Moscow.

Mathpal, Yashodhar (1984) *Prehistoric Rock Paintings of Bhimbetka*. Abhinav Publications, New Delhi.

Maturana, Humberto & Varela, Francisco (1980) *Autopoiesis and Cognition*. D. Reidel, Dordrecht.

Maturana, Humberto & Varela, Francisco (1998) *The Tree of Knowledge: Biological Roots of Human Understanding*. Shambhala, Boston.

McGilchrist, Iain (2009) *The Master and his Emissary: The Divided Brain and the Making of the Western World*. Yale University Press, New Haven.

McGilchrist, Iain (2021) *The Matter with Things: Our Brains, our Delusions, and the Unmaking of the World*. Perspectiva Press, London.

Michalak, Laurence & Trocki, Karen (2006) 'Alcohol and Islam: an overview', *Contemporary Drug Problems* 33:4, pp. 523–62. https://doi.org/10.1177/009145090603300401

Munsill, Lily (2014) 'Global ocean sprawl: a driver of jellyfish blooms' Wheaton College Research Depository: https://digitalrepository.wheatoncollege.edu/bitstream/handle/11040/23999/bio401_2015_munsill_lily.pdf (accessed 2022).

Murata, Sachiko & Chittick, William (2000) *The Vision of Islam*. I.B. Tauris, London.

Nagarjuna, trans. Garfield (1995) *The Fundamental Wisdom of the Middle Way: Nagarjuna's Mulamadhyamakakarika*. Oxford University Press, New York.

Ñanamoli, Bhikkhu & Bodhi, Bhikkhu (1995) *The Middle Length Discourses of the Buddha: A New Translation of the Majjhima Nikaya*. Wisdom Publications, Boston.

Newton, Roger (2004) *Galileo's Pendulum: From the Rhythm of Time to the Making of Matter*. Harvard University Press, Cambridge Mass.

Nietzsche, Friedrich, trans. Common (2006) *The Gay Science*. Dover Publications, Mineola New York.

Nussbaum, Martha (1994) *The Therapy of Desire*. Princeton University Press, Princeton.

Nyanaponika Thera and Bodhi, Bhikkhu (1999) *Numerical Discourses of the Buddha: An Anthology of Suttas from the Anguttara Nikaya*. Altamira Press, Walnut Creek California.

Özerem, Ayşem & Kavaz, Rahme (2013) 'Montessori approach in pre-school education and its effects', *The Online Journal of New Horizons in Education* 3:3, pp. 12–25.

Paganini, Gianni (2008) 'Montaigne, Estienne, et l'invention de l'apparence', *Philosophiques* 35:1, pp. 171–86. https://doi.org/10.7202/018244ar

Pakulski, Jan (2012) 'Introduction: John Higley's work on elite foundations in social theory and politics', *Historical Social Research* 17:1, pp. 9–20.

Park, Michael, Leahey, Erin, & Funk, Russell (2023) 'Papers and patents are becoming less disruptive over time', *Nature* 613, pp. 138–44. https://doi.org/10.1038/s41586-022-05543-x

Parkinson, C. Northcote (1957) *Parkinson's Law*. John Murray, London.

Perfetti, Charles & Goldman, Susan (1975) 'Discourse functions of thematization and topicalization', *Journal of Psycholinguistic Research* 4:3, pp. 257–71. https://doi.org/10.1007/BF01066930

Peterson, Kevin & Davidson, Eric (2000) 'Regulatory evolution and the origin of the bilaterians', *Proceedings of the National Academy of Sciences* 97:9, pp. 4330–3. https://doi.org/10.1073/pnas.97.9.4330

Piaget, Jean (1954) *The Construction of Reality in the Child*. Routledge, Abingdon Oxon.

Piaget, Jean (1955) 'The stages of intellectual development in childhood and adolescence' from Gruber & Vonèche, eds., *The Essential Piaget* (1977). Routledge, London.

Pinker, Steven (2011) *The Better Angels of our Nature: A History of Violence and Humanity*. Penguin, London.

Plato, trans. Cooper (1961) *The Collected Dialogues*. Pantheon Books.

Plato, trans. Waterfield (1987) *Theaetetus*. Penguin, London.

Polimeni, Joseph & Reiss, Jeffrey (2006) 'The first joke: exploring the evolutionary origins of humor', *Evolutionary Psychology* 4, pp. 347–66. https://doi.org/10.1177/147470490600400129

Popkin, Richard (1964) *The History of Scepticism from Erasmus to Descartes*. Van Gorcum, Assen Netherlands.

Popper, Karl (1963) *Conjectures and Refutations: The Growth of Scientific Knowledge*. Routledge, London.

Popper, Karl (1992) *Unended Quest: An Intellectual Autobiography*. Routledge, London.

Power, Camilla & Aiello, Leslie (1997) 'Female proto-symbolic strategies' from Hager, ed., *Women in Human Evolution*. Routledge, London.

Purser, Ron (2019) *McMindfulness: How Mindfulness became the New Capitalist Spirituality*. Repeater Books, London.

Putnam, Hilary (1982) 'Brains in a vat', from *Reason, Truth and History*. Cambridge University Press, Cambridge.

Rarick, Charles (2007) 'Confucius on management: understanding Chinese cultural values and managerial practices', *Journal of International Management Studies* 2:2.

Renqiang Li & 6 others (2014) 'Climate change-induced decline in bamboo habitats and species diversity: implications for giant panda conservation', *Diversity and Distributions: A Journal of Conservation Biogeography* 21:4, pp. 379-91. https://doi.org/10.1111/ddi.12284

Reza Husseini, Said, (2012) 'Destruction of Bamiyan buddhas: Taliban iconoclasm and Hazara response', *Himalayan and Central Asian Studies* 16:2, pp. 15-50.

Ricci, Jeanne (2018) 'The growing case for social media addiction' (California State University Blog). https://www.calstate.edu/csu-system/news/Pages/Social-Media-Addiction.aspx (accessed 2022).

Richardson, Anthony, Bakun, Andrew, Hays, Graham, & Gibbons, Mark (2009) 'The jellyfish joyride: causes, consequences, and management responses to a more gelatinous future', *Trends in Ecology and Evolution* 24:6, pp. 312-322. https://doi.org/10.1016/j.tree.2009.01.010

Rilke, Rainer Maria, trans. Spender & Leishman (1939) *Duino Elegies*. Norton, New York.

Robertson, Brian (2015) *Holacracy: The New Management System for a Rapidly Changing World*. Henry Holt, New York.

Rogers, Carl (1979) 'The foundations of the person-centred approach', *Education* 100:2, pp. 98-108.

Rogers, Lesley, Vallortigara, Giorgio, & Andrew, Richard (2013) *Divided Brains: The Biology and Behaviour of Brain Asymmetries*. Cambridge University Press, Cambridge.

Rohr, Richard (2016) *Breathing Under Water Companion Journal: Spirituality and the Twelve Steps*. SPCK, London.

Rosenthal, Seth (2006) 'Narcissism and leadership: a review and research agenda', *Center for Public Leadership Working Paper Series* 06-04. https://dspace.mit.edu/handle/1721.1/55948

Rossano, Matt (2006) 'The religious mind and the evolution of religion', *Review of General Psychology* 10:4, pp. 346-64. https://doi.org/10.1037/1089-2680.10.4.346

Roth, Gerhard (2013) *The Long Evolution of Brains and Minds*. Springer, Dordrecht.

Roth, V. Louise (1988) 'The biological basis of homology' from Humphries, ed., *Ontogeny and Systematics*. British Museum (Natural History), London.

Rubin, Jared (2012) 'Printing and Protestants: an empirical test of the role of printing in the Reformation', *Review of Economics and Statistics* 96:2, pp. 270-86. https://doi.org/10.1162/REST_a_00368

Russell, Bertrand (1952) *The Impact of Science on Society*. AMS Press, New York.

Russett, Bruce & Oneal, John (2001) *Triangulating Peace: Democracy, interdependence, and International Organizations*. Norton, New York.

Sandars, N.K. (trans., 1960) *The Epic of Gilgamesh*. Penguin, London.

Savoie, Donald (1994) *Thatcher, Reagan, Mulroney: In Search of a New Bureaucracy*. University of Pittsburgh Press. Pittsburgh.

Schlabach, Gerald (1996) *For the Joy Set before us: Ethics of Self-Denying Love in Augustinian Perspective*. University of Notre Dame ProQuest Dissertations Publishing 9621773.

Schmandt-Besserat, Denise (2006) *How Writing Came About*. University of Texas Press, Austin.

Schneiders, Antony & 5 others (2010) 'A valid and reliable clinical determination of footedness', *Physical Medicine and Rehabilitation* 2, pp. 835–41. https://doi.org/10.1016/j.pmrj.2010.06.004

Seligman, Martin (1995) 'The effectiveness of psychotherapy: the Consumer Reports study', *American Psychologist* 50:12, pp. 965–74. https://doi.org/10.1037/0003-066X.50.12.965

Sha, Zhiqiang & 9 others (2021) 'Handedness and its genetic influences are associated with structural asymmetries of the cerebral cortex in 31,864 individuals', *Proceedings of the National Academy of Sciences* 118:47, e2113095118. https://doi.org/10.1073/pnas.2113095118

Siniscalchi, Marcello & 5 others (2011) 'Sniffing with the right nostril: lateralization of response to odour stimuli by dogs', *Animal Behaviour* 92, pp. 399–404. https://doi.org/10.1016/j.anbehav.2011.05.020

Sjoman, N.E. (1996) *The Yoga Tradition of the Mysore Palace*. Abhinav Publications, New Delhi.

Smith, Allan H. (1960) 'The culture of Kabira, Southern Ryūkyū Islands', *Proceedings of the American Philosophical Society* 104:2, pp. 134–71.

Sogin, Mitchell & 7 others (2006) 'Microbial diversity in the deep sea and underexplored "rare biosphere"', *Proceedings of the National Academy of Sciences USA*, 103:32, pp. 12115–20. https://doi.org/10.1073/pnas.0605127103

Steiger, Jürg (2011) 'Why did mother nature provide us with two kidneys?', *Nephrology Dialysis Transplantation* 26:7, pp. 2076–8. https://doi.org/10.1093/ndt/gfr311

Stern, Daniel (1985) *The Interpersonal World of the Infant: A View from Psychoanalysis and Developmental Psychology*. Routledge, London.

Stern, Daniel (2010) *Forms of Vitality: Exploring Dynamic Experience in Psychology, the Arts, Psychotherapy, and Development*. Oxford University Press, Oxford.

Stolstorff, Melanie, Vartanian, Oshin, & Goel, Vinod (2012) 'Levels of conflict in reasoning modulate right lateral prefrontal cortex', *Brain Research* 1428, pp. 24–32. https://doi.org/10.1016/j.brainres.2011.05.045

Storm, Jason (2021) *Metamodernism: The Future of Theory*. University of Chicago Press, Chicago.

Swift, Jonathan (1729) 'A modest proposal'. Project Gutenberg: https://www.gutenberg.org/ebooks/1080 (accessed 2023).

Taleb, Nassim Nicholas (2012) *Antifragile: Things that Gain from Disorder*. Penguin, London.

Tatz, Mark (trans., 1994) *The Skill in Means Sutra*. Motilal Banarsidass, Delhi.

Taylor, John (2011) 'Prototype theory' from Meienborn, Heusinger & Portner, eds., *Semantics Theories*. De Gruyter, Mouton.

Teffer, Kate & Semendeferi, Katerina (2012) 'Human prefrontal cortex: Evolution, development, and pathology', ch. 9 from Hofman & Falk, eds., *Evolution of the Primate Brain*, pp. 191–218 (*Progress in Brain Research* vol. 195, Elsevier).

Treadgold, Warren (1997) *A History of the Byzantine State and Society*. Stanford University Press, Stanford.

Treisman, Jessica (2004) 'Coming to our senses', *Bioessays* 26:8, pp. 825–8. https://doi.org/10.1002/bies.20083

Trivellato, Francesca (2010) 'Renaissance Italy and the Muslim Mediterranean in recent historical work', *The Journal of Modern History* 82:1, pp. 127–155. https://doi.org/10.1086/650509

Trungpa, Chogyam (2002) *Cutting through Spiritual Materialism*. Shambhala, Boulder.

Uylings, Harry & Van Eden, Corbert (1991) 'Qualitative and quantitative comparison of the prefrontal cortex in rat and in primates, including humans', ch. 3 from Uylings et al., eds., *The Prefrontal: Its Structure, Function and Cortex Pathology*, pp. 31–62 (*Progress in Brain Research* vol. 85, Elsevier).

Vallortigara, Giorgio & Rogers, Lesley (2005) 'Survival with an asymmetrical brain: advantages and disadvantages of cerebral lateralization', *Behavioral and Brain Sciences* 28, pp. 575–633. https://doi.org/10.1017/S0140525X05000105

Van Valen, Leigh M. (1982) 'Homology and causes', *Journal of Morphology* 173:3, pp. 305–12. https://doi.org/10.1002/jmor.1051730307

Vaughan, Jill (2020) 'The ordinariness of translinguistics in Indigenous Australia', from Won Lee & Dovchin, eds., *Translinguistics: Negotiating Innovation and Ordinariness*. Routledge, London.

Visser, Arjan & Elena, Santiago (2007) 'The evolution of sex: empirical insights into the role of epistasis and drift', *Nature* 8: February 2007, pp. 139–49. https://doi.org/10.1038/nrg1985

Vivier, Eric & 7 others (2011) 'Innate or adaptive immunity? the example of natural killer cells', *Science* 331:6013, pp. 44–9. https://doi.org/10.1126/science.1198687

Wallis, Rijumati (2010) *Pilgrimage to Anywhere*. O-Books, Winchester.

Walshe, Maurice (1995) *The Long Discourses of the Buddha: A Translation of the Digha Nikaya*. Wisdom Publications, Boston.

Watford, Shelby & Warrington, Steven (2022) 'Bacterial DNA mutations'. National Library of Medicine: https://www.ncbi.nlm.nih.gov/books/NBK459274/

Weber, Max, trans. Parsons (1930) *The Protestant Ethic and the Spirit of Capitalism*. Routledge, London.

Weber, Max, trans. Henderson & Parsons (1947) *The Theory of Social and Economic Organization*. The Free Press, Glencoe Illinois.

Weil, Simone (2002) *Gravity and Grace*. Routledge, London.

Weil, Simone (2005) 'Human personality' from *An Anthology*. Penguin, London.

Westbrook, Raymond (2016) 'Jubilee laws', *Israel Law Review* 6:2, pp. 209–26. https://doi.org/10.1017/S0021223700002983

White, David (2012) 'Yoga, brief history of an idea', *Yoga in Practice* 5:1, pp. 1–23. https://doi.org/10.1515/9781400839933-004

Wightman, Gregory (2015) *The Origins of Religion in the Paleolithic*. Rowman and Littlefield, Lanham.

Wilber, Ken (1982) 'The pre/trans fallacy', *Journal of Humanistic Psychology* 22:2, pp. 5–43. https://doi.org/10.1177/0022167882222002

Wilber, Ken (2005) 'Introduction to integral theory and practice: IOS basic and the AQAL map', *AQAL: Journal of Integral Theory and Practice* 1:1, pp. 2–38.

Windsor, Prince Harry (2023) *Spare*. Bantam, New York.

Winnicott, Donald (1953) 'Transitional objects and transitional phenomena: a study of the first not-me possession', *International Journal of Psychoanalysis* 340, pp. 89–97.

Wisman, Jon & Smith, James (2011) 'Legitimating inequality: fooling most of the people all of the time', *American Journal of Economics and Sociology* 70:4, pp. 974–1013. https://doi.org/10.1111/j.1536-7150.2011.00795.x

Wittgenstein, Ludwig, trans. Paul & Anscombe (1969) *On Certainty*. Blackwell, Oxford.

Wordsworth, William, ed. Wordsworth (1995) *The Prelude: The Four Texts (1798, 1799, 1805, 1850)*. Penguin, London.

Xenophon, trans. Bowen (1998) *Symposium*. Aris and Phillips, Warminster.

Yucheng Guo, Pixiang Qiu, & Taoguang Liu (2014) 'Tai Ji Quan: an overview of its history, health benefits, and cultural value', *Journal of Sport and Health Science* 3:1, pp. 3–8. https://doi.org/10.1016/j.jshs.2013.10.004

Zaehner, R.C. (trans., 1969) *The Bhagavad Gita*. Oxford University Press.

Index

a priori, 168
Aborigines, see Australian Aborigines
absoluteness, 1, 4, 7, 14, 18–20, 22, 35, 79, 84, 98, 108–10, 128, 141–2, 144–5, 150, 151, 159, 161, 164–5, 167–9, 179, 189, 194–5, 199, 201, 205–6, 208, 228–31, 236, 252, 255, 272, 277, 281, 284, 287
absolutization, 4–6, 8, 11–12, 18, 20, **60–1**, 64–8, 70–1, 83, 95, 98, 115–16, 118–23, 125, 133–5, 140, 143, 146–8, 154–5, 157, 161–4, 170, 173, 179–80, 185–6, 188, 197, 199, 203–5, 209, 212, 229, 231, 256, 260, 275, 279–80, 282, 286, 292
Absolutization (book), 11, 60, 65, 74, 83, 105, 109, 116, 118, 121, 123, 125, 129, 131, 133, 135, 137, 139, 141, 143, 145, 147, 149, 151, 153, 155–6, 159, 161, 165, 167, 169, 171, 173, 175, 177–9, 181, 183, 185, 189, 191, 193, 195, 197, 199, 201, 203, 205, 207, 209, 235, 286, 290, 297
absolutized conflict, 25, 28, 65
absolutized specialization, 40
abstention, 150
abstract belief, 98, 113, 115–16, see also absolutization
abstraction, 157, 158, 178, 205–6, 243, 249
academic networks, 109
accelerating feedback loops, 14, 119, 127, 152, 183, 209
action, 6, 20, 37, **42–6**, 62–3, 82, 95, 184, 187, 198, 209, 232, 236, 247, 260–1

ad hoc argument, 116, 173
adaptability, 30, 131, 133, 187, 201, 268, see also adaptation; adaptivity
adaptation, 5–7, 15–16, 20–1, 23, **30–8**, 40, 45, 49, 53, 65, 102, 114, 118–19, 128, 136–7, 156, 163, 170, 184, 187–8, 199–200, 220, 261, 266–7, 284, 292, 294, see also adaptability; adaptivity
adaptivity/adaptiveness, 30–1, **32–7**, 38, 42, 49, 53, 111, 145, see also adaptability
addiction, 118, **147**, 150, 178, **180–1**, 184, 221, 223, 245
administration, 106, **194–203**
administrative reform, 202
administrators, 195, 201–3
adolescence, **93–9**, 213
adulthood, 6, 45, 70,87, 93, 95, **97–117**, 132
advertising, 162, 182, 244
aesthetics, 206, 217, **238–45**, 295
affordances, 82
Afghanistan, 162
Africa, 196, 214, 265, 278
agnosticism, 4, 6, 22, 67, **69**, 116, 122, 165, 167, 204, 230, see also even-handedness
agricultural revolution, 128
agricultural societies, 125, 132, 140
agriculture, 36, 45, 119, 122–5, **127–35**, 140–2, 149, 183, 189, 200, 216, 280
agroforestry, 127, see also forest gardens
alcohol, 149–50, 223, see also addiction
Alexander the Great, 165

alienation, 103, 114, 131, 192, 202, 228, 244–5, 254, 258, **274–6**
Amazonia, 130–1, 214
Ambedkarite Buddhism, 227
ambiguity, 16, 19, 37, 55, 67, 240, **259–62**, 301
amphibians, 52
analytic philosophy, 207, 246, **249–50**
anarchism, 194, 196–7
anarchy, 152
anatomy, 25
ancestor worship, 140
ancestry, 39, 51, 54, 68–9, 73, 128, 138, 140–1
Andean region, 155
Andrew, Richard, 47, 50–1, 56–8, 307
anima/animus, 193, 214, 217
animals, 13, 27, 29, 31, 36–7, 40, 42–52, 56–61, 63, 65–6, 70, 85, 87, 102, 114, 120, 127, 129, 131, 134, 140–1, 148, 170–2, 183–4, 187, 189, 197, 213, 223, 238, 276
anocracy, 9, 282
anthropology, 120, 170, 218, 259
antisemitism, 185, 262
Apartheid, 265
aphids, 31
apical organ, 50
Apocalypse, 9
appropriation, 118, 163, **167**, 204, 213, 216–17, 279, 282
Arabic, 254, 264
archaeology, 133, 137, 196
archetypes, 7, 84–5, 100, 105, 118, 136–7, 144, 146, 160, 172, 176, 193, 212, **213–17**, 225–6, 228, 234, 238–43, 263, 264, 282, 286–7
Archetypes in Religion and Beyond, 7, 85, 137, 216, 279, 301
Argentina, 140
Aristophanes, 259, 261
Aristotelian cosmology, 173
Aristotelianism, 248
Aristotle, 172–3, 248
arithmetic, 255
Ariyapariyesana Sutta, 264
Arjuna, 191
Arnhem Land, 271

arrow parable (of Buddha), 164
Arslantepe, 134, 280
art (visual), 33–4, 139–40, 160–2, 185, 214–17, **238–41**, see also cave art; images
arthropods, 48–9
artificial environments, 276–8, see also environment (immediate)
artists, 33–4, 244
arts, 3, 9, 33, 102–5, 204, 212, 217, 233, **238–45**, 254, 256, 274, 277, 286
Aryans (in India), 151
asanas, 235
asceticism, 118, 143, **147–51**, 162, 235, 234, 241, 268
asexual reproduction, 31
Asia, 227
Asian Buddhism, 230
assessment (educational), 252, 257, 301
association (psychological), 19, 62, 81, 83, 139, 143–4, 158, 215–16, 226, 239, 252–3, 259, 260
association (social), 283
astrology, 172
astronomy, 173, 255
asymmetrical integration, 78, 113
asymmetry (bilateral), 51, **52–9**, 60, 63, 113, 117, see also brain; hemispheres
asymptotes, 22
atheism, 108
Athens, 149, 206, 254, 280
attachment theory, 80–1
attention, 57–8, 62, 67–8, 82, 89, 110, 161, 166, 187, 217, 225, 237, **239–41**, 244, 263
attractive other, 213, see also anima
attunement, 81
Augustine, St, of Hippo, 264–5, 298
Australian Aborigines, 123, 259, 267, 271
Australopithecus, 52
Austria, 132
authenticity, 111
authoritarianism, 133, 137, 152, 155, 279, 283, see also power; totalitarianism

authority, 4, 86, 97, **99–100**, 101, 107, 116, 123–4, 133, 141–2, 144–5, 153, 159, 163, 165, 172–3, 177, 193–4, 196–9, 219, 222, 224, 241, 247, 261, 277, 279, 280, 294
autism, 88
autobiography, 211, **263–6**
autocatalysed, 293
autocatalysis, **14–15**, 20, 26, 27, 147, 293
autonomy, 37, 44, 197
autopoiesis, 12, 15
awareness, 22, 31, 55–6, 63, 66–9, 88, 90, 94–5, 98, 106, 110, 114, 120–1, 124, 128, 136–9, 143, 151, 158, 168–9, 175–6, 194, 208, 218, 223, 225, 227, 229, 233, 236, 237, 258, 262, 266, 268–9, 277–8, 282, 285–7, 290–2, 295
awe, 226, 278
Axial Age, 136, 142, 146, 219

Babylonia, 156
bacteria, 17, 30, see also single-celled organisms
balance, 25, 28, 76, 80, 83, 89, 110, 115, 193, 227, 229, 232, 235, 268, 277, 281–2, see also balancing; homeostasis
balanced effort, 229
balancing (action), 59, 131, 198–9, 214, 229, 232, **235**, 268, 282, see also balance; balancing feedback
balancing feedback, **5**, 6–7, 10–13, 16, 17, 19–20, 30, 32, 38, 41, 43, 45, 47, 60, 65, 71, 76, 88, 90, 95, 105, 113–14, 119, 133, 137, 156–7, 163, 169–70, 180, 184, 192, 212, 227, 229, 233, 236–9, 243, 245–6, 251–2, 258–9, 260, 262, 264, 267, 269, 277, 282–3, 286–7, 291–5, see also feedback loops; reinforcing feedback
Bali, 185
Bamiyan Buddhas, 162
Barandiaran, Xabier, 34, 44, 298
Beard, Mary, 260
beauty, 53, 58, **238–43**, 244, 278

Beck, Don, 77–8, 298
Beckwith, Christopher, 165–6, 298
Beethoven, Ludwig van, 217
behavioural addiction, 181, see also addiction
belief, 4, 7, 11, 16, **18–20**, 60, 63–4, 66, 69–70, 75, 80, 87, 89, 98, 105, 108–10, 115–16, 120, 136–40, 142–6, 149, 160–1, 165, 168–9, 173, 175, 194, 197, 205–6, 209, 213, 215, 219–20, 228–9, 231, 235, 249, 252, 255, 262, 267–8, 270, 286–7, 292
Bell-Lancaster system, 256
Bentham, Jeremy, 256
Bernard, St, of Clairvaux, 230
Bhagavad Gita, 190, 310
bias, 140–1, 170, 174–5, 180, 196, 246, 258, 286, 291, 295
Bible, 185, see also Hebrew ~ & under specific books
bilaterianism, 13, 38, 41, 46, **47–51**, 52, 66, 71, 113
bilingualism, 267, 270
binary division, 188, 204, 213, see also dichotomy; dualism
biodiversity, 127, 130, 183, 276
biology, 3–8, **10–71 passim**, 73–4, 94, 119, 174, 187–8, 197, 219, 283, 292
birds, 52, 57–8, 278
birth, 79, 80, 82, 89
body, 1, 15, 38, 40–2, 44–5, 47–56, 94, 150, 214, 225, 227, 229–30, 233, 237, 241, 255, 275, see also embodiment; body awareness; bodywork
body awareness, 227
bodywork, 9, **233–7**, 240, 254, 274
bonding, 82, 98–9, 138
Borges, Jorge Luis, 244
Bosnia, 1
Botticelli, Sandro, 242
bottom-up universality, 105, 117, see also universality; top-down universality
Bowlby, John, 81, 83–4, 298
Brahminism, 191, 226–7
brahmins, 190, 248

brain, 1, 5, 13–14, 19, 22, 28, 44, 46–7, **49–71**, 81, 114, 120–1, 168, 214, 229, 245, 293, see also brain lateralization; neocortex; parietal cortex; pre-frontal cortex; white matter
brain lateralization, 47, **49–64,** see also brain; hemispheres; left hemisphere; right hemisphere
Brazil, 124
breath control, 234
Brecon Beacons, 276
British Admiralty, 201
Buddha (Gautama), 9, 136, **143**, 147, 150–1, 163–6, 191, 206, 218, 220, 231–2, 246–8, 264–5, 268
Buddha's Middle Way, **150–1**, 165, 232
Buddha's scepticism, **163–6**, 206
Buddhism, 1, 4, 9, 141, 143–7, 150, 163–6, 191, 206, 216, 221, 226–31, 234, 236, 248, 254, 264, 268–9, 299, 304, see also Buddha; Buddhist tradition; enlightenment; Five Precepts; meditation
Buddhist tradition, 9, 146–7, 150, 164–5, 227–31
bureaucracy, 56, 102, 118, 142, 154–6, 194, 196, **199–203**, 272, see also administration; bureaucratization
bureaucratization, 106, 135, 257
burial, 125, 189
business, 56, 105–6, 109, 194, 198, 226, 258
Byzantine Empire, 161

calculus, 258
calendar, 172
California, 148, 306, see also US
caliphate, 145
calisthenics, 235
Calusa, 122
Canada, 123, 202
cannibalism, 125, 212, 214–15
capitalism, 149, 187, 191, 201, **274–6**
cartoon, 261

care, 92, 97, **99–101**, 107–8, 123, 141, 153–4, 190, 219, 222
carnism, 131
caste, 190–1
catalysis **12–15**, 26, see also autocatalysis
categorization, 52, 79, 170, 253, 270
Catholicism, 161–2, 167, 185, 225, see also Christianity; popes
causal beliefs, 138, see also causal theory; causal explanations
causal explanations, 286
causal processes, 2, 139
causal relationships, 170, 253
causal theory, 170, see also causal beliefs; causal explanations
caves, 138–40, 212, 214–16, 226, 238, 240
cave art, 139–40, 215, 226, 238, 240
cells, **12–15**, 17, 38–40, 42–4, 48
centralization, 133, 196, 199
cephalopods, 51–2
certainty, 1, 56, 60, 157, 161, 164, 168–9, 206, see also uncertainty
character (personal), 201, 265, see also role; virtue
character typology, 78
charisma, 154
charities, 194
chemical addiction, 180–1, see also addiction
chemical signals, 42, 43
Chen Wangting, 237
child, 14, 33, 49, 70, **81–93**, 98, 171,188–9, 214, 218–19, 223, 238–9, 253–7, 261, 274, see also childhood; education
child-centred learning, 257
childhood, 27, **81–93**, 211, 218–19, 286, see also child; adolescence
child-rearing, 96
China, 129, 142, 152, 157, 172, 200–1, 208, 233, 235–7, 241, 254, 264
Chinese classics, 201
Chinese religions, 216
Chomsky, Noam, 121, 302
Chrestomathia (Bentham), 256

Christ, 230, 269, see also Jesus
Christendom, 268
Christianity, 100, 132, 139, 144–6, 149, 162, 208, 216, 220, 223, 228–30, 241, 246, 249, 265, see also Catholicism; Christ; Christian Church; Orthodox Church; Protestantism
Christian Church, 109, 145, 173, 185
circular argument, 109, 116, 143, 246
cities, 127, 132–3, 191, 196, 277, 280, see also urban...
citizens' assemblies, 133
citizenship, 279, 281
civil servants, 177, 195
civil service, 177, 195, 200–2
class, 98, 101, 133, 143, 147–8, 152, 187, 190–1, 223, 226, 240, 254, 256, 272, 280
class struggle, 101
classical music, 243
classical mythology, 242
classics, 256
clerical workers, 254
clerics, 280, see also priests
climate change, 3, 45, 127, 204, 209, 284
clocks, 191
cnidarians, 43, 48
cognition, 18, 68, 75, 81–2, 87, 101, 240, see also belief
cognitive behavioural therapy, 288
cognitive development, 75
cognitive maps, 87
coherence, 21, 86
Cold War, 149
collective unconscious, 287
colleges, 257
colonialism, 122, 272, 281
comb-jellies, 42–3, 47
comedy, 240, 259–60
Comedy of Errors (Shakespeare), 260
commerce, 243, 272, 277, see also trade
commercialization, 211, 238, 243
communal tenure, 130
communication, 62, 98, 121, 162, 185–6, 208, 238, 272, 285

communications technology, 162, 185–6, 208, 272
Communism, 149, 153–4
community, 29, 38, 49, 86, 95, 176, 213, 279–80
competition, 5, 10, 25, 27–8, 45, 66, 94, 125, 145, 175, 298
competitiveness, 92
complexity, 1, 2, 4–8, 13, 19, 22–3, 27–9, 33, 37–40, 43–6, 49, 52, 54, 61, 65, 68–9, 86, 106, 114, 116, 125, 153, 156, 170, 179, 189, 252, 293–4, 303
comradeship, 98, see also friendship
concentration, 227–9
concepts, 1, 12, 16, 18–19, 21, 32–3, 55–6, 75, 94, 100, 102, 105, 120–1, 212, 216–17, 228, 231, 238–241, 244, 248, 256, 269, 274, 290
conceptual beauty, 244
concrete operational stage (Piaget), 91
conditioning, 73, 80, 114, 119, 152, 219
conditions, **4–6** and passim
Confessionsi (Augustine), 264–5, 298
confidence, 33, 75, 83, 84, 89, 99, 106, 115, 292
confirmation bias, 174, 286
conflict, 4, 15, **25–9**, 40, 64–6, 69, 70, 74, 76, 82, 91, 94–5, 98–9, 102–3, 109–10, 113, 125, 138, 146–7, 151, 161, 173, 181, 202, 207, 209, 218–21, 223, 228–9, 233, 235, 240, 247, 252, 254, 258, 266, 271, 272, 274–5, 276, 279, 281–2, 285–7
conformity, 40, 58, 197, 216, 221, 268–9, see also group
Confucianism, 201
Confucius, 143
Conklin, Beth, 214, 299
conscience, 93
consensus, 105, **133**, 196, 198, 204, 209, 267, 290, see also democracy
consequentialism, 107, see also utilitarianism
conservatism, **99–100**, 103, 124
conspiracy theories, 185–6

constitutions, 204, 281
contemplation, 225-6, 229
contemplative prayer, 225
contextualization, **4**, 59, 65, 68, 75, 76, 98, 102, 110, 117, 144, 215, 244, 266, 289, 291, 295
continuity, 9, 11, **12-14**, 39, 68, 74-6, 84-5, 94, 178, 237, see also discontinuity; incrementality
contracts, 95, 198, 275
conventionality, 287
cooperation, 63, 94, 130, 134
cooperative decision-making, 130, 134
Copernicus, Nicolaus, 173
copying, **38-41**, 59, 241
corpus callosum, 60, 62
corrupted, 283
corruption, 197, 200, 279, **283-4**
corvées, 133
cosmic justice, 149-50, 219
cosmology, 172, 228
cosmopolitanism, 123, 141, 267, 271
counter-culture, 99
counter-dependence, 147, see also unholy alliances
Cowen, Tyler, 174, 299
craving, 147, 162
creativity, 89, 94, 111, 192, 197, 207, 217, 230, 236, 241, 242, 245, 251, 274-5, see also balancing feedback
credibility assessment, 291
credibility factors, 258
critical thinking, 8-9, 103, 211, 255, 258, 284, **289-92**
Cromwell, Thomas, 191
cronyism, 283
crops, 129-31, 171-2, 182, see also agriculture
cross-wiring (of brain hemispheres), 47, 50-1
Crusades, 108, 230, 242
crustaceans, 48
ctenophores, 42-3, 48
Cueva Manos, 140
cults, 98

cultural development, 7, 61, 119, 293, see also culture
culture, 1, 3, 5-8, 59-61, 64-5, 73-4, 77-8, 81, 88, 91, 94, 99, 103, 111, 113-14, 116, 118, **119-292 passim**, 293
culture wars, 218
curriculum, 256
cybernetics, 179
cyclic history, 6
cycling, 262, 268
Cynicism (ancient), 248

dalits, 227
dance, 63, 83, 233, 237, **240**, 241
Daoism, 235
Darwin, Charles, 173
Darwinism, 32-6
Dawkins, Richard, 17, 299
Dawn of Everything, The (Graeber & Wengrow), 122, 302
death, 110, 125, 138-9
debt, 155-6, 183
debt forgiveness, 155
decision-making, 101, 133, 156, 198, 200, 202, 279-80, see also judgement
deduction, 1, 91, 97, 290, see also logic
deep generic skills, 257
defence, 12, 28, 100, 109, 168-9, 188, 282
deferring gratification, 149
deforestation, 183
deformation professionelle, 152-3, 275
defragmentation, 65
degenerative diseases, 132
deity, 228
delusion, 87, 147
democracy, 8-9, 111, 133, 145-6, 153-4, 177, 200, 216, 242, 267, **279-84**, 289
Democrats (US party), 101
denial, 150, 167
denialism, 150, 167, 169
deontology, 107
depression, 230, 245, 268, 285

Index

Descartes, René, 168, 250, 300
descendants, 37, 39
design, 39, 118, 178, 275
desire, 65, 67, 69, 75, 80-1, 86, 93, 118, **147-50**, 162, 178, 180, 193, 220, 281, 286
determinism, 6, 16-7, **18-21**, 28, 37, 44-5, 150, 176, 207, see also epistemic determinism
developing countries, 191
developmental psychology, 3, **72-117 passim**, see also stage theory
devotion, 143, 234
Dewey, John, 250, 283, 289, 300
dharma, 190-1
Di Paolo, Ezequiel, 34-5, 300
diachronicity, see time, awareness over
diagram, 263
dialectics, 111, 114, 167
dialogue, 167, 246-7, 250, 254, 267
Diamond, Jared, 29, 123, 132, 141, 150, 300
dichotomization, 186
dichotomy, 21, 122, 179, 192
Dickens, Charles, 256
differentiation, 1, 11, 13, 20, **80-2**, 131, 162, 206
digestive tract, 48
Diogenes Laertius, 165
direct democracy, 280, see also democracy
directionality, 21, 82
directiveness (in meditation), 225-32
disciplining (children), 218-19
discontinuity, 12-14, 16, 43, 69, 74, see also continuity; incrementality
discourse, 125, 171, 190, 248, 264, 267, 281
discussion, 39, 53, 75, 123, 163, 169, 248, 250, **257**
disembodiment, 192, 257, see also embodiment
disequilibrium, 75
DNA, 30-1, 39
dogma, 122-3, 139, 145, 164, 166, 168-9, 204, 213, 229, 263, 285, 297

dogmatic religion, 137
domain dependency, 103
dopamine, 151, 153, 181
double-blind testing, 174
drama, 240-1, 276
Dravidians, 151
dreams, 285-6
dualism, 115, 128, see also dichotomy
Dunbar number, 199
duration, 68
duty, 97-8, 107-8, 190-1, 218, 267, 274

early education, 257
Earth Overshoot Day, 183
East India Company, 201
echo chamber effect, 186
Eclogues (Virgil), 277
ecology, 26
Economic and Political Manuscripts (Marx), 274
economic development, 256
economic regulation, 142
economics, 62, 98, 103, 106, 135, 142, 145, 147, 149-50, 153-5, 175, 183, 191, 243, 256, 271, 274
eco-spirituality, 278
ecosystems, 3, 28, 45, 127, 183, 276
ecstatic experience, **138-9**, 215, 231, see also *jhana*; religious experience
education, 9, 33, 98, 102-3, 116, 145, 152, 154, 160, 174, 177, 193, 204, 208, 211, 222, 242, 246, 250-1, **252-8**, 272, 279, 281, 289-90, see also early education; higher education
egalitarianism, 119, **122-3**, 127-8, 130, 133, 148, 152-3, 194
Egypt, 172, 195, 220
Eifring, Halvor, 225, 227-9, 299-300
elections, 281
electrical charge, 11-13, **16-21**, 38, 42-3, 293
embeddedness, 84, see also embodiment
embodied meaning, 80-1, 253, see also meaning

embodiment, 3, 69, 80–1, 83, 110, 113, **114–15**, 118, 136, 143, 147, 152, 161, 164, 178, 205, **233–5**, 238, 242, 252–3, 256–7, 268, 272, 291
emergence, 6, **11–15**, 20, 25, 44, 84, 113, 153, 172
emotion, 55, **56**, 81, 83, 98, 101–3, 106, 109, 214–15, 227, 240, 242, 272, 288, 291
empathy, 55, 81, 91, **94–5**, 240
empiricism, 250
employment, 102, 182, 200, 202, 256, 275
Emptiness, 165
enactivism, 257
energy, 12, **25**, 28, 31, 39–40, 42, 44, 183–4, 234
Engels, Friedrich, 152, 196
engineering, 103
England, 136, 159, 191, 193, 256, 280
English Civil War, 280
English language, 270–2
enlightenment (nirvana), 70, 108, 113, 117, 138, 144, 165–6, 184, 265
Enlightenment (18th century) 238, 279, 281
enquiry, 9, 87, 109, 115, 211, **246–51**, 289, see also dialogue; philosophy
entoptic experiences, 139
environment, the (general) 59, 71, 114, 127–9, 132, 152, 178, **183–4**, 211, 270, 273
environment (immediate), 5–6, 12–13, 17–20, 25, 29, 31–2, 34–7, 40, 43, 45, 48, 63, 71, 76, 80, 82, 85, 89, 127–9, 132, 138, 151, 170–1, 178, 187–8, 195, 200, 237, 253, 257–8, 267, 270, **274–8**, 293, see also artificial environments; green environments; urban environments
environmental practice, 114
Epic of Gilgamesh, 158, 307
Epicureanism, 248
epidemics, 132
epigenesis, **73**, 114, 152–3, 187, 242
epiphanies, 141, see also insight; religious experience

epistemic determinism, 166, 207
epistemic humility, 167
epithelium, 43
equal opportunities, 192
equality (social), 123, **147–56**, 182, 190–1, see also egalitarianism
equality of justification, 204–5
equity, 223, see also justice
equivocation, 18
Erasmus, Desiderius, 167
Erikson, Erik, 75, 86, 301
error focus (in Middle Way Philosophy), 156
essay form, 244
Essays (Montaigne), 250
essentialism, 79, 117, 144, 248
eternalism (in Buddhism), 150
ethical observance, 9, 211, **218–24**
ethics/morality (no distinction intended), 9, 33, 75, 90, 92, 99, 105, 107, 115, 128, 137, 150, 166, 168, 191, 205–6, 211, **218–24**, 228, 237, 249, 261, 283, 294–5, see also principles, moral...
ethnic religion, 136, 141, 144
Etienne, Henri, 249
etymology, 271
eukaryotes, 25, 30–1, 37
Europe, 109, 122, 132, 137, 162, 191, 196, 233, 238, 241–2, 259, 272
Euthyphro, 247–8
eutrophication, 45
even-handedness, 111, 122, **163–7**, 176
evil, 144
evolution, 3–5, 15, 23, 26, 28, **30–7**, 40, 43, 47–9, 51–4, 56, 60, 68, 73, 74, 76, 85, 113, 157, 173, 187, 218, 281, 293
evolutionary development, 3, 36, 47
evolutionary selection, 40
examinations, 201, 257, see also assessment
executions, 195
Exodus, 142, 159, 161, 220
expanding circle (of sympathy), 223
expectations, 74, 88, 93–5, 97, 111, 182, 191

experientialism, 250
experts, 62, see also specialization
exploitation, 26, 46, 51, 56, 61, 118, 124, 127, 132, **147-52**, 153, 178, 180-2, 184, 223, 243, 273-4, 276
exponential growth, 61, 132, see also accelerating feedback
eyes, 17, **49-51**, 52-3, 57, 87, 139, 244

factories, 191, see also industrial revolution; manufacturing
facts, 92, 97, **207-9**, 252
fact-value dichotomy, 176
fallacies, 291, see also bias & under specific fallacies
false consensus, 175
false dichotomy, 21, 124, 176
false speech, 220
family, 84, 86, **88**, 89, 103, 107, 253, 258, 266
fantasies, 87, 89
farming, see agriculture
father, 89-90, 190, 247-8, 285, 291, see also parents
feedback loops, 5, 8, **11**, 27, 39, 50, 54, 65, 90, 119, 151, 157, 164, 184, 209, 212, 251, 261, 294, see also balancing feedback; reinforcing feedback
female genital mutilation, 107
femininity, 193
feminism, 190
fertility, 214
fidelity, 96
fighting, 233, 235, see also martial arts; violence
financial system, 183
fish, 29, 37, 45-8, 50, **52**, 58-9, 266
fishing, 124
fissiparousness, 146
Fitch, Tecumseh, 121, 302
fitness, 33, 53, see also adaptability; adaptivity
Five Precepts (Buddhism), 220
Five Principles (of Middle Way Philosophy), 4, 67, 69
Five Principles of Middle Way Philosophy, The (book), 8, 16, 22, 65-6, 68, 74-5, 109, 111, 113, 163, 167, 204, 229, 212-13, 231, 259, 291, 295, 297
flatworms, 47-9
flips, 58, 75, 78, 83, 108, 110-11, 145, 159, 208, 220, 222, 284
flood-retreat farming, 129
Florida, 122, 301
foetus, 82
folk art, 240, see also art
football, 95
footedness, 53
foraging societies, 124
Ford, Henry, 39
foreign language learning, 267, 270-2
Forest (in Buddha's life), 150-1
forest gardens, 130, see also agroforestry
forest monks, 230
formal education, 253-4, 257-8, see also education
formal operations (Piaget), 97, 101
formality, 252
formalization, 165, 212, 255, 285
foundational values, 99, see also absoluteness
foundationalism, 109, 143
Four Noble Truths (Buddhism), 166
fragility, 83, 127-8, 145-6, 150, 152, 155-6, 161, 181, 220, 224
framing, 2, **66**, 69, 71-2, 75-6, 92, 101, 107, 117, 135-6, 172, 176, 180, 202, 221, 224, 234, 247, 253, 263, 271, 275, 286, see also reframing
free association (in psychoanalysis), 285-6
freedom, 97, **99-101**, 107-8, **123**, 141, 145, 148, **153-5**, 158, 191, 197, 222, 269, 279-81
freewill, **6**, 16, 21, 28, 37, 42, 44, 150, 281
France, 124, 152
French language, 270
French Revolution, 152
Freud, Sigmund, 285-6, 288
friendship, 94, see also comradeship
frustration, 7, 147, 236
fungi, 226

Galileo Galilei,173
gallery, art, 245
games, 94, 274-5
Gandhara, 166
Ganges Valley, 166
ganglia, 49
gardens, 124, 129-30
Garstang, Walter, 36, 301
gender, 187-90, 193
General Studies, 193
generalists, 187
genes, 4, 23, **30-3**, 35-6, 39, 41, 48, 53, 73, 114, 187, 242, 253, 259
genetic mutation, 36
geology, 1, 173
geometry, 172, 179, 255
germ theory, 173
German language, 270
Germany, 132, 185, 243, 282
gerrymandering, 283
gestalt, 82, 241, 244
global warming, 183, 187, see also climate change
globalization, 204, 208, 211, 222, 256, 267, **272-3**
Glorious Revolution, 280
glyphs, 157-8
Gnosticism, 287
goal-orientation, 48, **56**, 268, see also goals
goals, 30, **34-5**, 48, **56**, 74, 82, 91, 95, 110-11, 137, 147-8, 177, 179, 198, 202, 213, 226, 231, 256, 268
God, 102, 136, 141-2, 144, 149, 159-61, 168, 215, 217, 220, 223, **225-6**, 228-30, 249, 268, 277
God's existence, 144, 249
goddess, 234, see also gods; anima
Godfrey-Smith, Peter, 13, 17, 38, 42-3, 45, 47, 51, 302-3
gods, 7, 136-7, 140-1, 144, 157, 172, **225-6**, 231
golden calf (Exodus story), 161
Goodall, Jane, 278, 302
gospels, 221
government, 56, 133-4, 136, 143, 165, **279**, 282-3, see also anocracy; democracy; kings

Graeber, David, 122, 124, 128, 130-4, 137, 148-9, 152, 155, 162, 189-90, 195, 219, 226, 279-80, 302
grain cultivation, 132
grammar, 158-9, 255-6, 272
graphemes, 158
gratitude, 225
grave goods, 138-40
Graves, Clare, 77, 302
gravity, 2
Greece, 20, 142, 163, 165-7, 206, 233, 240-2, 247, 254, 259-61, 272, 277, 298
Greek language, 272
green environments ('in nature'), 274, 276-8
green technology, 184
Griffiths, Mark, 181, 302
groundlessness, 115, see also uncertainty
group, 1, 12, 14, 27, 34, 40, 47, 50, 52, **58-9**, 60-1, 63-4, 72, 77, 88, 94, **97-104**, 105, 107-9, 114, 119, 123-4, 133-4, 137, 141-3, 146, 148-50, 153-5, 160, 172, 175, 185-6, 189, 194-5, 198, 208-9, 214, 218-22, 248-9, 253, 262, 264, 267-9, 279-81, 289
groupthink, 175
growth, 21, 23, 25, 41, 114, 132, 174, 183, 187, 256, 287
guilds, 196
gurus, 150, 165, 285
gymnastics, 233, 235, 254

Haeckel, Ernst, 36
Haidt, Jonathan, **99-100**, 107, 123, 153, 302
haikouella, 50, 52
hajj (Islam), 268
hallucinogens, 139
Hamlet (Shakespeare), 275
handedness, 53, 165
hands, 140
Hard Times (Dickens), 256
Harry, Prince, 266, 309
hatha yoga, 235, also see yoga
hatred, 147, 222, 275

Hauser, Marc, 121, 302
healing, 138, 172, 215, 262, 285, 288
health, 32-3, 49, 111, 132, 150, 204, 223
healthcare, 232
heaven, 149
Hebrew Bible, 141, 144, 159, see also Exodus; Leviticus; Joshua
hedonic treadmill, 180, 182, 231-2, 269
Heisenberg's Uncertainty Principle, 174
heliocentric theory, 173
Hellenistic philosophies, 248-9
hemispheres (of brain), 5, 13, **50-71**, 76, 82, 88, 90, 95, 104, 115, 120, 127, 131, 134, 140, 148, 155-6, 161, 173, 188, 197, 241, 290, 291, see also brain; left hemisphere; right hemisphere
Henning, Stanley, 236, 302
henotheism, 144
heroism, 9, 207, 215, 238-9, 264
Hibbing, Michael, 27, 302
hierarchization, 77-9, 122, 133
hierarchy of needs, Maslow's, 33
higher education, 102-3, 256, 258, 289, see also education; universities
hijacking (by absolutes), **119**, 134, 136-7, 143, 146, 157, 161, 163, 170, 176, 178-9, 195, 199, 203, 212-13, 251, 260, 262, 265, 292, 294
Hinduism, 9, 141, 162, 190-1, 216, 226, 228, 234, see also Brahminism; caste
historical narratives, 1
history, **1-9**, 11-12, 15, 32, 64, 78-80, 114, 116, 119-22, 128, 134-5, 137, 140, 144, 146-9, 152, 157, 161, 163, 167, 170, 178-9, 188-9, 192, 197, 209, 212-13, 215, 217-20, 223, 227, 230, 232, 236, 238, 246, 248-9, 254, 259, 293-4, 298
Hobbes, Thomas, 122, 250
hobbies, 274
holacracy, 198

holiness, 228, 247
holism, 13
Holy Spirit, 269
homeostasis, 25, 28, 184, see also balance
Homo erectus, 121
Homo sapiens, 52, 53, 121
homology, 10, **38-41**, 43, 47-8, 71
honour killing, 223
horticulture, 124, see also agriculture; gardens
Houellebecq, Michel, 26, 303
hox genes, 48
human nature, 117, 197
human rights, 102, 222
humani ties, 256
Hume, David, 168, 207, 246, 249-50, 303
humility, 230, 287
humour, 55, 211, 240, **259-62**
hunter-gatherers, 123, **124-31**, 141, 149, 170-1, 189, 214, 218, see also mesolithic; paleolithic
hunting, 28, 59, 129, 131, 140, 188-9, 214, 219
hypnotism, 285
hypocrisy, 222

iconoclasm, 157, 161-2, 216
iconophilia, 161-2
idealism, 150
idealization, 116, 122, 277
identity, 102, 141, 160, 220, 240, 260
ideological stage, 93, 97, **101-16**, 124, 153, 208, 218, 247, 253-4, 258
ideological thinking, 100-1, 103, 157, 169
ideology, 77-8, 93-4, **100-16**, 119, 131, 134, 136, 143, 146, 147, 148, 153-7, 160-1, 163, 169-70, 172, 174, 189, 202-3, 207-9, 218, 220-2, 231, 241, 243, 247, 253-4, 258, 267, 271, 277, 281, 284, 289, 294
idolatry, 118, 157, **159-62**, 185, 216
Iliad, 254
Illinois, 132

images, 55, 62, 106, 157, 159–62, 216, 228, 230, 234, 243, 255, see also art (visual); cave art
imagination, **18–19**, 97, 113, 115, 158–9, 178, 184, 238, 244, 285
imaging (technology), 54
imitation, 81, 241, see also copying
impatience, 56
imperial stage, **88–96**, 97–8, 100–1, 108, 201, 218–20, 253
imperialism, 271
impermanence, 227
impulse, 67, 81, 86–7, 89–90, 113, see also desire
impulsive stage, 80, **83–90**, 97, 253
'in over our heads', 74, 93, 100–1
Incas, 155
incorporative stage, **80–5**, 86, 90, 253
incrementality, 4, **16–17**, 20, 22–4, 35, 37, 39, 44, 67, **68–9**, 74, 110, 118, 204–5, 224, see also continuity
India, 142–3, 151, 162–3, 165–7, 172, 190–1, 201, 206, 219, 226–7, 233–6, 240, 242, 254, 268, 300
individualism, 95, 114
individuality, 88, 92, 111
indoctrination, 98
induction, 91
Indus Valley, 133, 196, 234
industrial revolution, 182, 191, 274, 277
inequality, 121, 147, 152, 154–6, 182, 190–1, see also equality
inevitability, 116, 143–4, see also determinism
infancy, 71, **80–5**, 86, 132, 219, 253
infanticide, 223
infinity, 70, 109, 116, 143, 161, 164, 231, 275
infinite rationalization, 109, 116, 143–4
inflation of logic, 255, 290
inflation of metaphysics, 109, 116, 143–4, 233, 235, 237, 252
information, 9, **16–19**, 22–3, 39, 42–4, 48, 50–1, 62, 66–7, 78–9, 84, 109, 157, 165, 173, 177, 182, 195, 198, 212, 279, 284, 291

ingroup bias, 175
inhibition, 61, 63
initiation ceremonies, 149, 196
injustice, 284, see also inequality; justice
insects, 48, 278
insecure attachment, 81, see also attachment theory; confidence
insight, 70, 78, 122, 147, 155, 160–2, 165, 168, 227, 231, 296
inspiration, 7, 84–5, 106, 137, 144, 146, 160, 172, 176, 211–12, **213–17**, 225, 228, 238–40, 258, 263–6, 277, see also archetypes
institutional stage (Kegan), see ideological stage
institutions, 99, 102, 105, 109, 111, 152, 170, 203, 280, 284, 294, see also administration
instrumentality, 52, **55–6**, 91–2, 116, 118, 127–8, 130–1, 140, 148–9, 151, 215, see also left hemisphere
integrated states, 71, 230, see also temporary integration
integration, **4**, 10–64 passim, **65–71**, 72–96 & 113–292 passim
integrative decision-making, 198
integrative practice, 211, **212–13**, 214–92 passim
intellectual integrity, 23
intelligence, 61, 68
intensive farming, 183, see also agriculture; instrumentality
intention, 82, 227, 229, see also free-will; judgement
intercultural contact, 143
interdependence, **8–9**, 13–14, 23, 33, 43–4, 52, 75, 122, 129–30, 148, 153, 163, 223, 229, 272, 282, 292
interdisciplinarity, **2**, 54, 193
interindividual stage, 6, 108, **109–17**, 136, 143, 146, 156, 169, 221, 284
intermediate goals, 9, see also goals
interpersonal stage, 78, 91–2, **94–104**, 107–8, 114, 116, 124, 136, 140–1, 143, 145, 153, 156–7, 161, 209, 218, 220, 222, 253, 267, 289, 294

interpretation, 35, 138, 140, 145, 154, 156, 159–60, 166, 172, 206, 221–2, 224, 235, 265, 286, 290–1
intoxicants, 220–1, see also addiction
intrinsic motivation, 92, 97, 257
inventions, 180
investment, 149, 152, 182–3, 231, 281
ions, 10, 12–13, 16–17, see also electrical charge
Islam, **144–6**, 150, 160, 162, 216, 241–2, 254, 268
Islamic law, 160
Islamic world, 145
islands, 129
Israelites, 141, 142, 155, 156, 161, see also Hebrew Bible
Italy, 173, 242
iterative homology, 10, **40–1**, 48
Itzkin, Elissa, 256, 303
Iyengar, B.K.S., 235

Jainism, 191, 220, 226
James, William, 250
Japan, 227, 236, 241, 264
Jaspers, Karl, 142, 219, 303
jellyfish, 43, 45–8
Jerusalem, 268
Jesus, 218, 221–2, see also Christ
jhana, 143, 225, 231–2
Johnson, Mark, 81–3, 238, 303
jokes, 260, 262, see also humour
Joshua, 142
journaling, 263–4
journeys, 260, 268–9
Joy, Melanie, 131, 303
Judaism, 141, 144, see also Hebrew Bible
judgement, **4–7**, 9, 21–2, 37, 42, 44, 50, 63, 65–6, 68, **73–9**, 81, 83, 89–90, 93–5, 99–100, 103, 106, 113, 115, 117, 119, 124–5, 128, 134, 136–7, 140, 142–6, 148, 156, 160, 162–3, 166, 169, 172, 174–5, 180, 186, 188, 197, 202, 205, 207–9, 212–13, 221, 225, 232, 247, 257–8, 265, 267, 272, 282–3, 289, 290, 292–3, 295

judgement focus (in Middle Way Philosophy), 156
Julius Caesar, 264
Jung, Carl, 193, 285–7, 303
Jungian psychotherapy, 287
Jungianism, 287
Just War, 222
justice, 97, 99, **100**, 102, 107–8, 114, 123, 141, **153–5**, 222, 280–1, see also equality
justification, 22, **109–10**, 115, 139, 143, **166–70**, 205–7, 209, 235, 246, 252, 290–2, 296
Juvenal, 277, 303

Kandiaronk, 123
Kant, Immanuel, 107, 168, 250
Kantianism, 115
karma, 150, 164, 219, 221
Kegan, Robert, 6, 72–4, **76–9**, 80–117 passim, 124, 253, 303
kidneys, 41, 49, 54
killing, 28, 191, 220, 223
kindness, 222
kings, 133, 190, 195–6, 275, 280, see also government
kinship, 195, 218
kitsch, 243
Klikauer, Thomas, 203, 304
knowledge, 6, **18–20**, 164, 168, 171, 176, 196, 201, 206, 228, 252, 255–8, 270, see also belief; justification; truth
knowledge by acquaintance, 270
Kohlberg, Lawrence, 75, 86, 92, 115, 304
Konner, Melvin, 218–19, 304
Krishna, 190
Krishnamacharya, Tirumalai, 234–5
Kropotkin, Peter, 196–8, 304
kshatriyas, 190–1
Kuhn, Steven, 188–9, 304
Kuhn, Thomas, 76, 304
kundalini, 234

labour, 130, 132–3, 149, 182, 191, 275, see also work

Lahontan, 124
Lakoff, George, 190, 304
lancelet, 50-2, 54, 56
land, 47, 49, 52, 125, 127-30, 183, 189, 195, 200, 220, 271
landscape, 87, 242-3, 265
language, 34, 55, **61**, 64, 80, 85-7, 100, 108, 110, 119, **120-1**, 128, 135, 158, 170, 211, 238-9, 253, 255, 264, 267, 269-73, see also language acquisition; foreign language learning; propositional language
language acquisition (children), 85-7
Larkin, Philip, 30, 304
Laszlo, Ervin, 87, 304
lateralization of brain, see brain lateralization; hemsipheres
Latin (language), 51, 249, 271-2
laughter, 259-60 see also humour
law, 142, 144, 159, 200, 219, 248, **279-81**, see also Islamic law; law (Jewish)
law (Jewish), 160, 222
lay people, 220
leadership, 29, 93, **99**, 123, 142, 154, 195, 198, 231, see also authority
learning, 4, 76, 88-9, 93, 97, 106, 157, 158-9, 170, 172, 176, 211, 245, **252-8**, 264-5, 267, 270-3, see also balancing feedback; education
learning difficulties, 93
left hemisphere (of brain), 50, **54-64**, 66, 68-9, 88, 91, 120, 131, 161, 194, 197, 199-200, 214, 241, 244, 295, see also hemispheres (of brain)
left-right axis, 47
legal framework, 195
legal norms, 199
legalism, 160
leisure, 192, 274, see also recreation
leukocytes, 40
Lévi-Strauss, Claude, 124, 171-2, 304
Leviticus, 141, 155, 159
Lewis, Bernard, 145, 304
liberalism, 32, 78, 100-1, 107
liberal arts, 255

liberalization, 222
liberation, 145, 147-8, 153-4, 156, 163
liberation of women, 145
libertarianism, 154-5, 194
Liebermann, Philip, 121, 304
life (living systems), 3, 5, 10, **11-17**, 19-21, 25, 27-8, 30, 32, 38-9, 49-50, 56, 60, 71, 113, 184, 212, 238, 278, 292-3
life (individual life span), 79-80, 82, 87, 109-10, 119, 150, 218, 263-6
life-story, 264
linearity, 1-2, 4
linguistics, 3, see also language, meaning
listening, 239, 243, 257, 283
literacy, 118, **157-60**, 253-4, 256
literature, 158, 165, 217, **238-43**
livestock, 131, 157, see also carnism; meat
local government, 282
Locke, John, 122, 250
locomotion, 48, see also movement
logic, 91, 97, 211, 255, **290-2**, see also deduction; inflation of logic
logical positivism, 250
Long Walk to Freedom (Mandela), 265, 305
looking, 51, 54, 105, **239**, 243
Louisiana, 195
love, 100, 110, 221-3, 230
loyalty, 91, 97, **99-101**, 107, 116, 141, 153-4, 222, 238
lumping, 163, **167**, 168, 204
lute strings analogy (Buddha), 232
Lutheran Reformation, 145
lying (false speech), 90, 220

MacIntyre, Alasdair, 95, 305
madrasahs, 254
magic, 172, 215, 234
Magna Carta, 280
Maharaja of Mysore, 235
Mahayana Buddhism, 165
Maiden of Yue, 236
maladaptiveness, 7, 25, 38, 266
Malunkyaputta (Pali Canon), 164
mammals, 36-7, 52

management, 103, 194, 196, 203
managerialism, 106, 118, 194, 203
Mandela, Nelson, 265, 305
Manichaeism, 265
mantras, 228
manual dexterity, 253
manufacturing, 183-4, 275, see also factories; industrial revolution
Marcus Aurelius, 264, 305
market, 102, 182, 198
market forces, 102
marriage, 96, 106, 205
martial arts, 233, **235-7**, 254
Marx, Karl, 1, 152, 196, 274-5, 305
Marxianism, 182
Marxism, 146
masculinity, 193
Maslow, Abraham, 33-4
massage, 233
materialism, 176
materials, 26, 180, 183, 245, 275
mathematics, 61, 172, 176, 240, 244
Mato Grosso, 124
Matthew (gospel of), 221-2
Maturana, Umberto, 12, 17-18, 174, 305
maturity, 95, see also developmental psychology
McGilchrist, Iain, 53-5, 60-3, 67, 69, 81, 88, 115, 161, 174, 238, 241-2, 244, 265, 291, 305
Mcmindfulness, 230
meaning, 1, 16, **18-20**, 55-6, 65, 68, 72, 75, **80-7**, 89, 110, 121, 125, 136, 139, 142, 158-9, 162, 172, 185, 199, 222, 233-5, 239-40, 243, 252, 253-6, 258-64, 267, 270-2, 286-7, 295
meaning-making, 80
measurement, 172
meat, 131, 184, 223, see also carnism; livestock
Mecca, 268
mechanism, 22
mechanistic biology, 6
media, 102, 185, 262, 283
medical research, 174
medicine, 132, 285

meditation, 3, 8-9, 143, 165, 211, **225-32**, 233, 236, 276
Meditations (Descartes), 168, 250, 300
Meditations (Marcus Aurelius), 264, 305
meiosis, 31
melody, 238
memoir, 264-5
memorization, 257, 264, 272
memory palace, 264
men, 188-9, 193, 214, 223, 271
menstrual blood, 214
mental health, 228
mental illness, 60, 87, 286, 291
mental states, 67, 125, 221, 225-6
Mesoamerica, 129, 130, 137, 157
mesolithic period, 124-5, 141, 279
Mesopotamia, 129, 132-3, 172, 254, see also Babylonia
metabolism, 15, 23, 44
metamodernism, 250
metaphor, 55, 161, 190, 232, 253, 268, 270
metaphorical extension, 253
metaphysical inflation, see inflation of metaphysics
metaphysics, 6, 9, 33, 37, 64, 83, 92, 105, 108-9, 115-16, 117, 128, 136, 138, **143-4**, 145-6, 149-50, 159, 164-6, 168-9, 176, 206, 227-8, 233, 235, 237, **246-51**, 254-5, 269, 277, 287, 292
metazoa, see multicellular organisms
Mexico, 133
Michelangelo Buonarroti, 242
Middle East, 129, 142, 162, 196
Middle Way, passim, especially **4-5**, **65-71**, **143**
Middle Way Philosophy (explicit mentions), 1-2, 13, 34, 77, 252, 267, 294-7
mid-life crisis, 110
migration, 208
military service, 102
mind-body system, 3, 115
mindfulness, 9, 225, 229, 230, 232, see also meditation
minimal state, 202

mirror neurons, 81
missionaries, 106
mitigated scepticism, 168, 207
mobility, 45, 53, 106, 125, 190–1
models (conceptual), 17, **58–9**, 72, **76**, 86, 91, 105, 117, 131, 137, 178, 194, 228, 241, 253–4, 267, 270, 286–7
modern art, 244
modernism, 250
modesty, 148
Mohenjodaro, 133, 196
Moldova, 195
molluscs, 47
monarchism, 143, see also kings
monasticism, 141, 216, 230, 236, 254
Monet, Claude, 242
money, 152, 180
monism, 115
monolingualism, 271
monotheism, 144, 157, 160
Montaigne, Michel de, 167, 250
Montessori, Maria, 257
moral absolutism, 150
moral conventionalism, 166, see also moral relativism
moral development, 75, 107, 115
moral motives, 218
moral observance, 218–24
moral principles, 75, 115, see also principles
moral relativism, 150, see also moral conventionalism; relativism
moral rules, 92, 219, see also rules
morality, see ethics (no distinction intended)
Moreno, Alvaro, 34, 44, 298
morphology, 25
mortality, 132, see also death
Moses, 142, 159, 161, see also Exodus
mother, 49, 80, 81, 114, 190, 239, 253, see also parents
motivations, 218, see also desire; goals
Mount Sinai, 142, 159, see also Exodus

movement (locomotion), 21, 23, **25**, 42–4, 47–8, 50, 58–9, 68, 82, 184, 233–7, 263
movement (social change), 98–9, 166, 208, 217, 227, 243, 277, 287, 290
Muhammad, 1, see also Islam
Mulamadhyamakakarika (Nagarjuna), 165, 305
Mulroney, Brian, 202
multicellular organisms (metazoa), 13, 15, 17, 30–2, **38–41**, 42–5, 47–9, 113
multidisciplinarity, see interdisciplinarity
multi-word stage (language acquisition), 85
muscles, 42–5, 53, 56, 187
music, 55, 83, 99, 211, 217, **238–43**, 245, 254–5, 278
musical instrument, 245
Muslims, see Islam
mutual aid, 196–8
mutual gaze, 81–2
Mycenae, 272
myelin sheathing, 61
mysticism, 146, 161, 225, 229–30
mythology, 161, 216, 242

Nagarjuna, 165, 206, 305
Nambikwara, 124
Napoleon, 152
narcissism, 93
narrative, see story
Narwari, 141
Natchez, 195
nation, 51, 190, see also state
Native Americans, 77, 122–3, 130, 132, 148, 195, 267
natural killer cells, 40
natural resources, 122
naturalism, 204–9
nature, 21, 128, 168, 176, 207, 216, 274, **276–8**
Nazism, 185, 243
needs, 3, 6, 25–6, 28, **33–6**, 38, 41, 46, 48–9, 57, 86, 93, 95, 110, 114,

129–30, 134, 155, 182, 187, 198, 200, 253, 257, 274, 282
negative absolutization, 204
neocortex, 61, 70, 88, see also brain
neoliberalism, 202
neolithic period, 7, 64, 122, 124–5, 127, 131–2, 139–40
neomania, 180
Nepal, 141
nepotism, 283
nervous system, 10, 17–19, 37, **42–5**, 50, 52
nested systems, 2–3, 119
network, 109, 198
neural links, 1, 5, 19, 68, 82, 260, 267, 270, 293
neural network, 11, 18, see also brain; nervous system; neural links
neuroscience, 3, 121, 291, see also brain
neurosis, 286, 288
New Guinea, 123, 141, 196
Newton, Isaac, 2, 173
Newtonian physics, 173
niche, evolutionary, 26–7, 31, 33, 36–7, 118, 187–8, 293
Nietzsche, Friedrich, 207, 263, 305
nihilism, 115, 150, 206–7
nirvana, see enlightenment
nirvana fallacy, 108, 184
nisharum, 156
noble savage, 122, 126
non-conceptuality, 229
non-directive meditation, 228–9
non-violence, 222
normalization, 118, 178, 180, 184, 190
normativity, 34
North America, 129, 279
Northwest Coast (of North America), 148
no-till cultivation, 127
novel, 26, 57, 243, 245, 256
numeracy, 253
nurturing mother metaphor, 190
Nussbaum, Martha, 249, 305

O'Keeffe, Georgia, 243

obedience, 86, 102, 197, 219, 221
observation, 54, 57, **170–4**, 201–2, 248, 257
obsessive-compulsive disorder, 60
Odyssey, 254
Ohio, 132
old age, 79
Oneal, John, 281, 307
ontogeny, 5, 36, 73, 85, 293
ontology, 5, see also metaphysics
oppression, 155, 191, see also repression
optionality, 16, **21–4**, 61, 68, 75, 80, 85, 131, 153, 179, 184–92, 195, 207, 213, 221–2, 231, 245
oral poetry, 239
organic farming, 127, see also agriculture
organisms, 5, **10–64 passim**, 114, 147, 184, 293
organizations, 92, 102, 111, 132–3, 183, 190, **194–203**, 280, 282, 293, see also administration
organs (bodily), 38, 40, 45
original language fallacy, 255
Orthodox Church, 162
Ottoman Empire, 145
Outlines of Pyrrhonism (Sextus Empiricus), 249
over-conceptualization (of arts), 238, 243–4
over-consumption, 149
over-specialization, 170, **187–93**, 244, 274–5
over-technologization (of arts), 238, 243–4
ovulation, 214
Oxford Dictionary of Philosophy, 205, 298

Pacific, 129
Pacifism, 222
pain, 230, see also suffering
painting, 226, 240, 242–3, see also art (visual)
Palace (in Buddha's life), 150–1
paleolithic period, 7, 64, **119–26**, 133, 136, 138–41, 146, 163, 170, 189,

214-15, 219, 226, 238, 267, 271, 279, 285
paleoneurology, 69
Pali Canon, 164-5, 231, 248, 254, 264-5
panopticon, 256
paradigm, 76, 124, 174
paradigm shifts, 76
parasitism, 30
parental discipline, 218
parents, 30-1, **81-7**, 90, 92-4, 97-9, 106, 214, 218, 253-4, see also father; mother.
parietal cortex, 65, 69, see also brain
Paris Climate Agreement, 209
Parkinson, Northcote, 201-2, 306
Parkinson's Law, 202, 306
parliaments, 279-81, see also democracy
parochialism, 102, 105, 107, 141
particularity, 118, 204-5
parties (political), 98
pastoral literature, 277
pasture, 130
Patanjali, 227, 234
path, 9, 11, 20, **23-4**, 37, 82, 85, 94, 221, 232, **268**
pathology, 54
patriarchy, 187, 189, 192
peasants, 191, 294
pedagogy, 255, see also education
peer relationships, 258
peer review system, 175, 251
peers, 59, 94, 99-100, 143, 258
peregrini, 269
perennial food plants, 130
Perfection of Wisdom (Buddhism), 165
performing arts, 238
periodic table, 258
permaculture, 127
Permian period, 47
Persia, 142, 200
persons, 78-9, 149
perspective, **2-4**, 33, 66, 75-6, 87-8, 92, 98, 109, 113, 117, 134, 139-40, 166, 205, 207, 215, 235, 242, 244-5, 251, 266, 283, 294

petitionary prayer, 225, 228
Petrarch, Francesco, 242
phenomena, 11, 16, 90
phenomenology, 246, 250
Philippines, 129
Philogelos, 260
philosophical education, 250
philosophy, 2-3, 9, 21-2, 78-9, 116-18, 142, 144, 146, 163, 165-9, 176, 181, 192-3, 204-8, 242, **246-51**, 252, 254, 264, 274, 281, 283, 285, 289-90, 294-6
philosophy of science, 176
phonetics, 158
phonological system, 121
phylogeny, 5, 36-7, 40-1, 85, 113-14, 119, 293
physical contact, 81, 219
physicalism, 176
physicians, 172
Piaget, Jean, 6, 74-5, 86, 91, 97, 101, 158, 306
pilgrimage, 267-70
Pilgrimage to Anywhere (Wallis), 269, 309
Pinker, Steven, 125, 223, 243, 272-3, 282, 306
Plains Indians, 123
planning laws, 200
plants, 20, 26-7, 31, 42, 45, 130, 139, 170-2, 187, 215, 226, 276
Plato, 166-7, 206, 247-8, 250, 254-5, 306
Platonism, 117, 287
play (free activity), 89, 93, 238, 250, 257, 275
plays (drama), 240, 260
pleiotropy, 39
poetry, 158, **239**, 243-5, 254, 263-4, 269
political administration, 177
political correctness, 224
political philosophy, 197
politics, 2, 8, 61, 63, 65, 67, 79, 97-101, 103, 105-7, 109-10, 116, 121-5, 128, 133, 144, 146, 151, 153, 155-6, 165, 168, 177, 182, 189-90, 193-4, 197, 200-4, 218, 222, 224,

226, 231, 247, 270–2, 277, **279–84**, 294–5, see also democracy; government; ideology
pollution, 183, 270
polygenesis, 39
Polynesia, 129
popes, 94, 173, 277
Pope Benedict XVI, 94
Popper, Karl, 37, 286, 306
populism, 262
portraiture, 242
positive absolutization, 204–5, see also absolutization
Possibility of an Island, The, 26, 303
postmodernism, 108, 250, see also relativism
potlatch ceremonies, 148
poverty, 261, see also equality
power (socio-political), 1, 73, 77, **78–9**, 93, 103, 121, 123, 136, 138, 141, 152, 182, 189–91, 195, 201, 203, 227, 257, 259–60, 271–2, 275, **279–84**, 294
practical religion, 137
practice, 2–3, 5, **8–9**, 11, 32, 60, 65, 67, 69, 71, 73–4, 76, 78–80, 89, 91, 95, 100, 109–10, 116, 120, 137, 140, 143–6, 151, 156–9, 163–6, 169–71, 176–7, 179–180, 182, 187, 189, 197–8, 203, 205–6, **211–95 passim**
Practice of Agnosticism, The, 22, 297
pragmatism, 246, 250
praise, 225
prayer, 211, **225–6**, 228–30, 232
precepts, 220–1, see also ethical observance; principles
predation, 10, **25–9**, 40, 42, 45, 49, 52, 56–7, 59, 66
predictability, 19, 22, 45, 58
prediction, 172
pre-frontal cortex, **60–9**, 229
Prelude, The (Wordsworth), 265, 310
pre-operational stage (Piaget), 86
prescription, 161, 220
Pre-Socratic philosophy, 247
pre-trans fallacy, 71, 80
priests, 127, 172, 190, 196, 216
primates, 52, 61, 68, 70, 85

primitive communism, 122, 196
principle of reticence, 79
principles, 75, 103, 107, 115, see also ethics; Five Principles; Five Precepts; rules
printing, 185
probabilities, 22, 39, 179
probabilizing, 22
procrastination, 201
production (economic), 127, 130, 181–2, 184, **274–5**
production line, 39
professions, 105, 190, 223, 253, 289
professionalism, 98, 105–6, 187, 190, 223
professionalization, 285
profit, 182, 203
progressive politics, 105
prohibition of alcohol, 150
projection, 7, 103, 118, **136–7**, 140–6, 172, 176, 212–13, 215–16, 219, 223–5, 229, 235, 243, 274, 277, 282, 285
proliferation (of mental states), 63, 164, 227, 229, see also reinforcing feedback
propaganda, 162, 185, 238, 243, 265
property, 4, 12, 122, 148, 155, 195, 223, 281
prophets, 222
proportionality, 93, 111, 182, 241
propositional language, 64, 72, 85–6, 110, 120, 121, 253
propositional markers, 64
propositional thinking, 69, see also representationalism
propositions, 91, 97, 142, 157, 160, 176, 253
Protagoras, 206
Protestantism, 149, 161–2, 185
Proteus, 20
provisional farming, 127, 128
provisionality, 2, 4, 6, 11, 16, **18–21**, 55, 63, **67–9**, 73–4, 76–7, 79, 83, 102, 106, 117–18, 120–5, 127–8, 131, 134, 137–8, 140, 142–3, 147, 154, 156–8, 161, 163, 166, 168–9, 173–80, 184, 186–191, 194–5,

197–200, 203–5, 212, 218, 220–2, 228, 242, 263, 267, 271, 279, 280, 292, 295
prudence, 220
pseudoscience, 172, see also scientism
psychoactive substances, 215, 226
psychoanalysis, 78
psychology, 3, 5–7, 44, 72–117 passim, 119, 123, 131, 134, 153, 158, 194, 197, 202–3, 218, 219, 223, 228, 252–3, 274–6, 283, 291, 293–4
psychosis, 286
psychotherapy, 193, 211, **285–8**
public policy, 256
publication bias, 175
punctuated equilibrium, 6, see also tipping points
punishment, 86, **218–20**, 256
puns, 260
purity, 97, **99**, 101, 107, 116, 141, 153, 179, 222, 226
Putnam, Hilary, 168, 306
Pyrrho of Elis, 165
Pyrrhonism, 163, **165–6**, 248–50, see also scepticism

qigong, 237
quadrivium, 255
qualifications (academic), 257
quality of life, 132
quantum physics, 174
Qur'an, 160, 254

race, 25, 43, 49, 223, 262, 281
racial segregation, 281, see also Apartheid
racism, 262
raiding, 124
rare biosphere, 27
rational consistency, 115
rationalism, 168, 250, 290, see also Platonism
rationality, 194, 199–200, 216, 290
reading, 125, 254, see also literature
Reagan, Ronald, 202
realism, 150, 176

reality, 20, 61, 64, 91, 143, 168, 200, 250
reason, 13, 21, 32–3, 39, 59, 61, 77, 111, 115, 164, 168, 176, 180, 184, 186, 226, 281, 283, 290–1, see also reasoning
reasoning, 86, 91, 100–2, 105, 168, 289, 291–2, see also deduction; logic
reassurance, 81, 211, 259–261
rebellion, 123
rebirth, 164, 221
recitation, 255
records (written), 194, 199, 203, 236, 247, 264, see also administration
recreation, 187, 192, 211, 233, 268, 274–6
recursion, 121, 265
Red Book (Jung), 286
red ochre, 139, 212, 214–15
redistribution, 130, 152, see also justice
reductionism, 136, 138–9, 218, 286
redundancy, 47, 49
reflection, 2, 79, 160, 163, 167, **218–22**, 254, **263–5**, 288, 295
Reformation (Protestant), 161, 185
reformations, 145
reframing, 65, **66**, 69, 71, 117, 194, 236, 240, 292, see also framing
reinforcing feedback, **5–8**, **11 passim**, see also balancing feedback; feedback loops
relational thinking, 2, 69
relationships, personal, 33, 56, 73, 75, 84, 91, **94–5**, 98, 105, **106**, 137, 197, 199, 214–15, 221, 223, 238, 258, 285–6, see also friendship; parents; sexual relationships
relative values, 100
relativism, 78, 118, **204–10**, 218
relativization, 243–4
relaxation, 227, 229–30, 235
religion, 1–4, 7–8, 78, 105–6, 108, 110, 116, 135, **136–46**, 159–62, 170, 172–3, 176, 185, 211–17, 219, 225–6, 229–31, 238, 242–3, 250, 255, 259, 268, 287

Index   331

religious experience, 138, 140-3, 146, 211-12, 216-17, 238, see also insight; inspiration; *jhana*
religious tradition, 1, 4, 137, 139, 146, see also religion
Renaissance (European), 168, 173, 191, 217, 238, 242, 277
renewable energy, 184
renewable resources, 183
Renoir, Pierre-Auguste, 217
renunciants, 216, 226
representation, **18-19**, **61**, 64, 85, 110, 118, **120-1**, 157-8, 161, 176, 179, 229, 257, 281, 292, see also meaning; belief
representationalism, 18, 176, see also meaning; representation
repression, 1, 25, 27-8, **60-4**, 66, 92, 106, 118, 120-1, 123, 131, 134, 149, 151-2, 162, 192-3, 197, 220, 243, 245, 259, 265, 286
reproduction, 6, 10, 21, 23, **30-3**, 35, 37-41, 71, 113, see also sexual reproduction
reptiles, 52
republics, 133-4, 137, 143, see also democracy
Republican (US party), 101, 155
resilience, 27, 83, 268.
responsibility, 8, **21-2**, 81, 93, 98, 137, 178, 197, 202, 219, 221, 224-5, 228, 247
responsiveness, 15, 38, 42, 45, 71
re-thematization, 157
retreats, 107
revelation, 9, 144, **231**, 235
revolution, 147, **152-4**, see also agricultural ~; French ~; industrial ~; Russian ~
reward, 92, 149, 150, 182, 219-20, 256
rewilding, 127
rhetoric, 202-3, 254-5
rhythm, 238-9
right hemisphere (of brain), 50-1, 53, 56, 58, 63, 65-6, 88, 95, 120, 140, 148, 197, 214, 241, 245, 265, 291, see also brain; hemispheres
rigid beliefs, 61

Rijumati Wallis, 269, 309
Rilke, Rainer Maria, 70, 307
rites of passage, 95-6, 271
ritual, 99, 123, **138-40**, 149, 195, 215-16, 226, 233, 238, 240
Robertson, Brian, 198, 307
Rogers, Carl, 285, 287, 307
Rogers, Lesley, 47, 50-1, 56-8, 307, 309
Rohr, Richard, 225-6, 307
role (social), 86, **88-9**, **92**, 96-7, 190-1, 198-9
Romans, 166, 242, 248, 260, 262, 264, 272, 277
Romanticism, 217, 243-4, 265, 277
Rome, 240, 259, 268, 277, 280
Rosenthal, Seth, 93, 307
Rossano, Matt, 138-9, 214-15, 307
rote memorization, 252, 254
Rothko, Mark, 243-4
Rousseau, Jean-Jacques, 22
rulers, see kings
rules, 86, 90-1, **92**, 97, 140-1, 144, 149-50, 152, 155, 159-60, 194-5, 197, 199, 200, 203, 205-6, 211, 219-22, 271, 279, 281
rumination, 227, 263, see also proliferation; reinforcing feedback
Russell, Bertrand, 172-3, 307
Russett, Bruce, 281, 307
Russia, 152, 282, see also Soviet Union
Russian Revolution, 152
Ryukyu Islands, 170

sacredness, 161, 226, 228
sacred text, 228
samatha meditation (Buddhism), 227-9
samurai, 227
San, 214
Sanjaya Bellatthaputta, 164-5, 206
Sanskrit, 240
sassatavada, 150
satire, 261-2
Savoie, Donald, 202, 307
Scandinavia, 153
scarcity value, 182

sceptical argument, 105, 109–10, 161, **163–9**, 174, 176, 205–6, 247, 249–50, see also scepticism; uncertainty
sceptical enquiry, 111
sceptical slippage, 204
scepticism, 4, 30, **32**, **67–9**, 111, 115, 135, 163–9, 170, 176, 205–6, 246, see also sceptical argument; uncertainty
schemas, 253, 268
schismogenesis, **148**, 150–1, 162, 219, 226
schisms, 145
schizophrenia, 54, 244, 291
scholarship, 2, 4, 221
school, 88–9, 92, 94, 154, 232, **252–8**, see also education
school phobia, 89
science, 2–3, 8–9, 11, 76, 78, 105, 134, 142, **170–7**, 178–9, 181, 192, 208–9, 242, 248–9, 252, 256, 258, 277, 289, see also scientific method; scientism
science of the concrete (Lévi-Strauss), 171–2, 178
scientific method, 134–5, 170, 173, 179, 258, 289
scientism, 106, 108, 118, 170–3, 175, 179, 192
scriptures, 1, 143–5
Scythia, 166
seasonal shifts, 124
Second World War, 161, 282
secure attachment, 80–1, 83
security, 33, 81–3, 85, 89, 283, 295
self-actualization, 33–4
self-denial, 223
self-discipline, 148
self-indulgence, 148, 151
self-mortification, see asceticism
self-organization, **11–15**, 16, 19–21, 38
self-view, 6
semantic system, 121
Semendeferi, Katerina, 68, 308
sensations, 239, see also senses
sense-organs, 48

senses, 16–7, **43–4**, 45, 48–9, 71, 82, 109
sensitivity, 17, 23, 38, 43, 92, 293
sensori–motor patterns, 18, 28, see also meaning
sentences, 85–6, 158, see also propositions
sequence, 3, 44, 56, 71, 243
serfdom, 190–1, see also slavery
serial monogamy, 106, 111
Sermon on the Mount (Jesus), 221
sex (activity), 30, 32–3, 171, 188, 205, see also sexual relationships; sexual reproduction
sex (difference), 171, 188, see also gender
Sextus Empiricus, 166–7, 249
sexual desire, 286
sexual maturity, 94
sexual misconduct, 220
sexual relationships, 98, 214–15, see also relationships; sex (activity)
sexual reproduction, **30–3**, 37, 113
sexuality, 219
Shakespeare, William, 242, 260
Shakti, 234
shamanic healing, 138, 140, 211
shamanism, 138, 140, 211, 215–16, 285
Shi'i Islam, 162
shirk (Islam), 160, 162
shortcuts, 1–2, 8, 12, 22, 53, 63, 79, 137, 140, 142, 146, 153, 198, 216, 243, 280
shramana, 248
shravaka, 226
shudras, 190
Shukallituda, 171
silence, 164
single-celled organisms, 11, 17, 19, 21–3, 25–6, 28, 30, 38, 43, 46, 114
skilful means (Buddhism), 144
skilled workers, 182, 256
skills, 9, 95, 103, 120–1, 124, 131, 182, 235, 239, 252–4, 256–7, **258**, 289, 290–2, see also communication; critical thinking; foreign language learning

slapstick, 259, 260, 262
slaughter, 131, see also livestock; meat
slavery, 148, **151-2**, 155, 182, 190, 220, 223, 281
social codes, see rules
social function, 213, 215
social groupings, see group
social media, 67, 98, 181, 185, 186, 262, 269, 283
social order, 221, see also class
social proof, 175
social relationships, see relationships, personal
social sciences, 138, 193, see also anthropology; psychology
socialism, 101
society, **3**, 88, 97-9, 106, 119, **121-6**, 130-1, 141, 147-8, 152-4, 171, 189, 192, 198, 203, 209, 219-20, 240, 243, 253, 275, 279, 283, 293
sociopathy, 93
Socrates, 143, 166-7, 206, 246-8, 250
Socratic questioning, 166, 247
solar system, 173
solidarity, 63, 97, 130, 134, 138, 197, 213, 215-16, see also group
Sonadanda (Pali Canon), 248
Sophism, 206
South Africa, 214, 265
Southwood, Ben, 174, 299
sovereignty, 194-7
Soviet Union, 154, 243, see also Russia
space, 68, 82, 135, 157, 249, 275
Spare (Prince Harry), 266
Sparta, 149
specialism, 127, 175-6, 181, 216, see also specialization
specialization, 10, 38, 40, 50-2, 54, 61-2, 92, 113, 129, 135, 172-3, 175, 178, 180-1, **187-93**, 212, 214, 233, 238, 243-4, 249
species, **3-5**, 15, **23**, 26, 30-1, **33**, 37, 56, 59, 114, 120, 131, 171, 187, 293-4
speculation, 9, 116, 165, 246
Spiral Dynamics, 77-8

spiritual teachers, 143, 150, 235, 248, see also gurus; teaching
sponge, 45
sport, 233, 274-5
spouse, 97, see also relationships, personal
Spring and Autumn Annals, 236
St Peter's Square (Rome), 243
stage theory, **73-9**, 80, 99, also passim 72-117
stage transition, 6, 83-5, 88-90, 94-6, 98, 101-4, 109-13, 145, 253, 258
stages (of human development), 5-6, 11, 17, 23, **72-117**, 119, 214, 218, 220, 252-3, 258, see also incorporative, impulsive, imperial, interpersonal, ideological, & interindividual stages respectively
state, 17, 119, 194, 201, 203, 283
state societies, 125
stealing, 90, 220
Stern, Daniel, 82, 308
stern father model, 190
Stiner, Mary, 188-9, 304
Stoicism, 248-9, 264
story, 2-7, 15, 19, 52, 55, 60-1, 64-5, 73, 79, 93, 115, 127, 147, 150, 158-9, 188, 236, **239-40**, 250, 263-5
storytelling, 239-40
straw-manning, 206
stress, 8, 31, 111, 225, 230, 237, 276, 283
stretch (moral), 26, 74, **107**, 218, 224, 234-5, 240, 260, 272
stretching (bodywork) 234-5
structural coupling (Maturana & Varela), 17-18
student-centred learning, 252, 257
studying, 270
sub-atomic particles, 3
subjectivism, 204, 206-7, 209, see also relativism
sublimity, 162, 217, 242, 274, 278
sub-optimization, 4, see also adaptivity

substantive freedom, 123, see also freedom
substitution, 2, 155, 269
suffering, 214, 240, 281, 285
Sulawesi, 130
Sumeria, 133-4, 142, 157, 171, 190, 196, 280, 282
sunk costs fallacy, 180
Sunni Islam, 162, 216
supernatural, 7, 12, 136, 137-8, 140, 160-1, 213, 215, 219, 225-6, 228, see also supernaturalism
supernaturalism, 211, 217, see also metaphysics; naturalism
supervision, 194
suppression, 60, **62-3**, 67, 92, 120-1, 220
surpluses, 127, 129, 132
surveillance, 256
survival, 6, 23, 30, 33-6, 57, 58
survivorship bias, 246
sustainability, 7, 47, 106, 109, 128, **130-1**, 137, 149, 152, 156, 183-4, 215
Swift, Jonathan, 261, 308
symbiosis, 30
symbolism, 82, 85, 95, 139, 150, 162, 211, **212-17**, 228, 238, 239, 240, 243, 244, 264, 287
symbols, 1, 7, 56, 75, 84-5, 137, 139-41, **212-17**, 226, 228, 238, 240-1, 252, see also archetypes
symmetry, 47-8, 52-3, 62, see also asymmetry (bilateral)
Symposium (Xenophon), 254, 310
syncretism, 230
synthesis, 233, 236
system of equivalence, 155
systemic conflict, **25-8**, 65, 70
systemic history, 2-3
systemic interpretation, 258
systemic specialization, 40
systems, **2-8**, 11-18, 21, 25, 28, 34, 37-8, 42-3, 50, 62, 106-7, 114, 128, 153, 155, 174, 195, 198, 283, 291
systems biology, 3, 174
systems thinking, see systems

taboo, 9, 29, 215, 260
tai chi, 237
Tai Ji Quan, 236-7
Taiwan, 129
Taleb, Nassim Nicholas, 179-80, 308
Taliban, 162
tanha, 147
Tantra, 234
Taosi, 152
task-positive network, 229
tawhid (Islam), 144
taxation, 152, 280
teacher training, 257
teaching, 238, 247, **252-8**, 290, 292, see also education; spiritual teachers
team ethos, 95
technology, 8, 114, 132, 135, 142, 152, 170, 176, **178-86**, 189, 192, 208, 225, 232, 245, 256, 267
Teffer, Kate, 68, 308
telegraphic stage (language acquisition), 85
teleology, 34
temples, 190, 216
temporary integration, 231, 233, 276, see also integration; *jhana*
Teotihuacan, 133
Thatcher, Margaret, 202
Theaetetus (Plato), 206, 306
theatre, 240-1
theism, 176, 208
thematic meditation, 228-9
thematization, 158-9
theocracy, 165
theology, 144, 221
theory, **2**, 11-13, 18, 73-9, 87, 97, 99, 115, 124, 128, 151, 171, 173, 176, 178-9, 211, 221, 234, 250, 255, 263, 275, 281, 286-7, 291, 293
theory of mind, 87
therapy, 230, 233, 249, 287, see also psychotherapy
Theravada Buddhism, 230
thermodynamics, 25
threat responses, 137
threats, 7, 12, 43, 56, 100, 213, 215, 220, 238-9, 282

Three Gorges Dam, 208
Tibet, 165, 241
time, see history; time allocation; time, awareness over
time, awareness over (diachronicity), 4, 7–9, 32, 36, 40, 56, **63–71**, 85, 87, 137, 157–8, 207, 263
time allocation, 191–2, 201–2
tipping points, **16**, 37, 43–5, 73, **74–6**, 77, 80–1, 84, 88, 90, 113, 115, 117, 153, 177, 184, 253, 293
Tlaxcala, 133, 137, 280
toddlers, 80–1, see also infancy
Tom and Jerry (cartoon), 260
tone, 55, 238
top-down authority, 194, 197, 199, 202, see also power
top-down universality, 105, see also universality
Torah, 141, 159–160, 220, see also Exodus; Hebrew Bible; Leviticus
torture, 125
totalitarianism, 282, see also authoritarianism; power
totemism, 139
tourism, 269–270
toxicity, 149, see also intoxicants
tracking (continuous attention), 2, 68
traction, 47
trade, 127, 129, 132, 142–3, 157, 190, 272
traditions, **1–2**, 8, 33, 105, 129, 140, 151, 166, 170, 190–1, 216, 219, 222, 226–31, 233–7, 240–1, 243, 248, 254–5, 264, 267–8, 277, see also Buddhist tradition
traditional societies, **99–100**, 124
tragedy, 240
training, 74, 206, 211, 220, 235, 236–7, 252, 257, see also education
training principle, see precepts
transition, see stage transition; also see tipping points
transitional object, 84
translingualism, 267, 271–2
transport, 181, 267
trauma, 27, 215

travel, 44, 48, 123, 141, 185, 208, 211, 264, **267–73**
trial and error, 178–9
triumphalism, 108
trivium, 255
Trump, Donald, 94, 209, 284
trust, **68**, **94–5**, 113, 171, 177, 199, 263, 279, 281–2
trustworthiness, **68**, 201
truth, 168, **204–9**, 216, 227, 255
Turkey, 134

ucchedavada, 150
UK, 65, 202, 254
Ukraine, 133, 166, 195, 282
uncertainty, 4, 30, 34, 67, 109–10, 118, 163–5, 167, 169, 213, 269, see also scepticism
unconscious, 162, 286–7
understanding, 62, 88, 97, 115, 136, 144, 158–9, 171, 173, 195, 224, 238, **252**, 253–5, 261, 267, 273, 283, see also meaning
unholy alliances, 204
unions (workers'), 196
universality, 34, 72, 74–5, 77, 97, **100–6**, 107–9, 111, 114, 124, 133, 136, 140–1, 143–4, 146, 152–4, 156, 159–60, 169, 176, 191, 207, 222, 248, 256, 281, 289
universe, 3, 164, 173, 178, 228, 231
universities, 73, 97, 102, 255
Upanishadic tradition, 226–7
urban civilizations, 127, 189–90
urban environments, 274, **276–8**
urbanism, 196, 216, see also urban civilizations
Uruk, 133
US, 101, 149–50, 155, 202, 209, 227, 281, 290, see also under specific states
utilitarianism, 33, 78, 204, 208, 255–6, 261, 283

vaccination, 171
vagueness, 16, 21, 22
vaishyas, 190

Vallortigara, Giorgio, 47, 50–1, 56–8, 307, 309
value foundations, 97, 99, 101
values, 1, 6, 11, 32–3, 36, 40, 57, 66, 71, 75, 78, 82, 92–5, 97, **99–102**, 111, 116, 123–4, 140, 142, 145–6, 148, 151, 153–6, 160–3, 168–9, 206–9, 217–18, 231, 222, 233, 235, 244, 249, 264, 272, 281, see also aesthetics; ethics/morality
Van Valen, Leigh, 38–9, 309
Varela, Francisco, 12, 17–18, 174, 305
variation, 10, 39–41, 58, 122, 125
varna, 190
Vatican, 243, see also popes
Vaughan, Jill, 271, 309
Vedanta, 144
Vedic Hinduism, 144
veganism, 66, 127, 249, see also carnism; meat
vegetarianism, 28, 249, see also carnism; meat
Venice, 280
ventral-dorsal axis, 47
Venus, 214
vertebrates, 48, 50, 51–3, 57, 63, 70
vested interests, 283
viability, 35
village monks (Buddhism), 230
violence, **125**, 133, 190, 195, 220, 222, 223, 260–2, see also war
vipassana meditation (Buddhism), 227–9
Virgil, 277
Virgin Mary, 230, 243
virtue, 75, 95–6, 107, 150, 201, 228
visual processing, 57, 139, see also eyes; images; art (visual)
vitality effects, 82
Vivekananda, Swami, 234
vocabulary, 55, 270, 272, 283
vocal tract, 120–1
vocational courses, 103
volunteering, 270
voting, 8, 101, 198, 282–3
Vygotsky, Lev, 74

Wales, 276

walking, 47, 184, 263, 268, 276
war, 1, 97, 125, 146, 196, 281, see also violence
Wari', 214
warriors, 134, 151, 190
wealth, 135–6, 149, 152–3, 181, 183, 190
Weber, Max, 149, 199, 309
Weil, Simone, 161, 309
welfare state, 152, 191
Wendat, 123, 279
Wengrow, David, 122, 124, 128, 130–4, 137, 148–9, 152, 155, 162, 189–90, 195, 219, 226, 279–80, 302
West, the, 98, 101–2, 108, 145, 153, 162, 167, 173, 201, 208, 223, 230, 233, 236–8, 240, 242–3, 249, 256, 271–2, 281
Western Buddhists, 145, 230
White, David, 234–5, 309
white matter (brain), 68
Whitman, Walt, 269
Wightman, Gregory, 136, 309
Wilber, Ken, 71, 77, 309
wilful effort, 229, 231
Winnicott, Donald, 84–5, 310
wisdom, 68, 201, 221, 230
Wittgenstein, Ludwig, 169, 310
wokeness, 224
Wolsey, Cardinal, 191
womb, 70, 80–1, 114
women, 172, 188–9, 193, 214, 223, 280–1
words, 1, 18, 75, 82, 85, 152, 158–9, 238–9, 255, see also language
Wordsworth, William, 265, 310
work, 94–6, 103, **110–11**, 133, 148, 182, 191–4, 201–2, 270, **274–6**
working class, 101, 193
wrestling, 235
writing, 142, **157–9**, 160–3, 172, 185, 219, 239, 254, 263–4
writing systems, 157
Wyatt, Thomas, 242

Xenophon, 254, 310

Yahweh, 141, 156, 159

Yangtze, 208
Year of Jubilee, 155–6
yoga, 233–7
Yoga Sutras, 227, 234
Yucheng Guo, 237, 310

zazen, 227
Zhang Senfeng, 237
Zoroaster, 143
Zoroastrianism, 166
zygote, 114

www.ingramcontent.com/pod-product-compliance
Lightning Source LLC
Chambersburg PA
CBHW071954220426
43662CB00009B/1129